Innovations in Process Engineering

Held at the 2024 AIChE Annual Meeting

San Diego, California, USA
27-31 October 2024

ISBN: 979-8-3313-1686-0

Innovations in Process Engineering

Held at the 2024 AIChE Annual Meeting

San Diego, California, USA
27-31 October 2024

AIChE
The Global Home of Chemical Engineers

Printed from e-media with permission by:

Curran Associates, Inc.
57 Morehouse Lane
Red Hook, NY 12571

Printed with permission by Curran Associates, Inc. (2025)

For permission requests, please contact AIChE
at the address below.

AIChE
120 Wall Street, FL 23
New York, NY 10005-4020

Phone: (800) 242-4363
Fax: (203) 775-5177

www.aiche.org

Additional copies of this publication are available from:

Curran Associates, Inc.
57 Morehouse Lane
Red Hook, NY 12571 USA
Phone: 845-758-0400
Fax: 845-758-2633
Email: curran@proceedings.com
Web: www.proceedings.com

TABLE OF CONTENTS

BREAKTHROUGHS IN SUSTAINABLE CHEMICAL PRODUCTION AND PROCESS

CONVERSION OF PLASTIC WASTES TO HIGH VALUE PRODUCTS

EFFICIENT PROCESSING OF LIGNIN TO BIOPRODUCTS AND BIOFUELS

INTEGRATED PROCESS ENGINEERING AND ECONOMIC ANALYSIS

IONIC LIQUIDS: NOVEL SEPARATION, CATALYTIC REACTION AND ELECTROCHEMICAL PROCESSES

MATERIALS FOR THERMOCHEMICAL AND ELECTROCHEMICAL ENERGY STORAGE

NEW PROCESSES FOR EFFICIENT CO2 CAPTURE AND UTILIZATION UNDER MILD CONDITIONS

PROCESS INTENSIFICATION AND MODULAR MANUFACTURING: INTENSIFIED REACTION AND SEPARATION PROCESSES

TOPICAL PLENARY: FRONTIERS IN GREEN PROCESS AND PRODUCT ENGINEERING (INVITED TALKS)

329a A Nature-Inspired Approach to Design Hierarchical Zeolites As Efficient Catalysts for Bio-Syngas to Olefins Conversion

Nidhi Kapil, Mohammed Babkoor, Massimiliano Materazzi, Marc-Olivier Coppens

The global production of light olefins, including ethylene, propylene and butylene, is substantial, amounting to more than 250 million metric tons annually, as they serve as building block for various products, including polymers, cosmetics, rubbers, and detergents.[1] The conventional method to produce olefins involves steam cracking of natural gas or naphtha, which consumes a lot of energy and relies on non-renewable resources. Therefore, demand has increased for sustainable pathways that address resource limitations and reduce environmental impacts. In this regard, the catalytic conversion of syngas into olefins is desirable, and several technologies have been developed to improve this route including Fischer-Tropsch synthesis and methanol from syngas, followed by Methanol-to-Olefins (MTO). This could make the overall process more sustainable, if syngas is produced from a renewable resource, like biomass or waste.

A nature-inspired solution methodology will be employed, taking inspiration from the structures of leaves to synthesise hierarchical zeolites that will facilitate direct conversion of syngas to olefins, synergistically with a methanol synthesis catalyst.[2] The common issue encountered with the reported catalysts is deactivation by coke accumulation, hence a strategy to enhance the catalyst lifetime is required. Therefore, zeolites with precisely controlled meso- and macropores will be synthesised to control molecular diffusion and mitigate the effects of catalyst deactivation.[3] Furthermore, the modulation of surface barriers for effective control of mass transfer will be achieved through treatments of zeolites, utilising methods such as acid etching and SiO_2 deposition.[4] The catalysts will be characterised using multiple spectroscopic technique such as NMR, gas physisorption, NH_3-TPD and HRTEM. The catalytic activity of the prepared catalyst for the direct syngas to olefins will be discussed.

[1] S. Zhao, H. Li, B. Wang, X. Yang, Y. Peng, H. Du, Y. Zhang, D. Han, Z. Li, *Fuel* **2022**, *321*, 124124.
[2] M. -O. Coppens, T. Weissenberger, Q. Zhang, G. Ye, *Advanced Materials Interfaces* **2021**, *8*, 2001409.
[3] T. Weissenberger, N. Kapil, P. Trogadas, M. -O. Coppens, *ChemCatChem* **2022**, *14*, e202200268.
[4] S. Xu, K. Zheng, C.-R. Boruntea, D.-G. Cheng, F. Chen, G. Ye, X. Zhou, M.-O. Coppens, *Chem. Soc. Rev.* **2023**, *52*, 3991-4005.

329b Sustainable Synthesis of C2-Building Blocks Via Ethanol Oxidative Dehydrogenation on Iron-Molybdenum Mixed Oxide Catalysts

Danny Stark, Alfons Drochner, Bastian Etzold

Introduction

Modern chemical industry is increasingly emphasizing highly efficient and sustainable processes. Rather than relying on crude oil, coal and natural gas as primary feedstocks, a rapidly growing number of waste reduction, reutilization, and circular strategies is being engineered. While not being exclusively motivated by the goals of defossilization, many sustainability approaches incorporate the usage of bio-based raw materials for the development of drop-in technologies. One chemical that is already available in large quantities and produced based on renewable resources is ethanol. Currently over 100 billion litres of bioethanol are produced annually by the fermentation of starch- and sugar-based feedstocks alone.[1] Hence, a key objective of current catalysis research is dedicated to the development of ethanol upgrading processes for the sustainable production of platform chemicals.[2] One possible route for the utilization of ethanol is its conversion to acetaldehyde via oxidative dehydrogenation (ODH), a route analogous to the well-established *Formox* process used in industry for the large-scale production of formaldehyde from methanol. This heterogeneously catalyzed process utilizes iron molybdate catalysts. Remarkably, only few studies are available regarding the applicability of these catalysts for the conversion of ethanol to acetaldehyde, even though this route provides a sustainable alternative to the currently applied Wacker-Hoechst process for acetaldehyde production via liquid phase ethylene oxidation.[3] Therefore, this study evaluates the iron molybdenum mixed oxide system as a prospective catalyst for the ODH of ethanol to acetaldehyde.

Materials and Methods

Iron molybdenum mixed oxide catalysts of varying composition were prepared by coprecipitation of their respective metal salts. Performance and stability of these catalysts were investigated in a continuous flow fixed bed tubular reactor. Steady state as well as transient response methods were applied for kinetic investigations. Off-gas was analyzed by a quadrupole mass spectrometer and an online gas chromatograph. Structural characterization of catalysts was done by N_2-Physisorption, NH_3-TPD, ICP-OES, Raman spectroscopy, XRD, SEM and XPS.

Results and Discussion

The evaluation of catalysts with varying compositions highlights an interesting finding: while pure iron oxide promotes the combustion reaction of ethanol, even the introduction of small molybdenum quantities provides catalysts with excellent acetaldehyde selectivity, allowing for single pass yields exceeding 90 %. Furthermore, these iron rich catalysts demonstrate superior activity compared to the $Fe_2(MoO_4)_3$ benchmark and enable almost full ethanol conversion at low temperatures while maintaining high selectivity (refer to Figure 1a). Alongside achieving equivalent yields at lower reaction temperatures, the use of iron-enriched catalysts has the additional benefit of extending the catalysts lifetime.

Our research demonstrates that the deactivation of molybdenum oxide-based catalysts in the ODH of ethanol proceeds via the depletion of active material by volatilization, analogous to the deactivation when employing methanol as the substrate.[4] However, the deactivation is kinetically hindered at lower temperatures allowing the highly active iron rich catalysts to retain their activity (refer to Figure 1b), while the benchmark shows decreasing ethanol conversion and acetaldehyde selectivity over time. We will delve into the immense changes in both performance and stability among catalysts with different compositions, considering disparities in their structural and surface composition, as corroborated by a range of characterization techniques; i.e. we were able to correlate differences in product selectivities with differences in specific surface acidity, as evidenced by NH_3-TPD experiments (refer to Figure 1c).

Finally, steady state kinetics for the ethanol ODH, kinetics of bulk oxygen diffusion and desorption kinetics were studied for our highly active iron-rich catalyst. Through variation of feed composition and investigations into the consecutive oxidation of acetaldehyde and acetic acid kinetic parameters were obtained. Furthermore, we employed transient response methodologies to unveil novel mechanistic insights. Specifically, the involvement of bulk oxygen in the reaction was validated through pulse experiments (refer to Figure 1d). Lastly, a deeper understanding of surface reaction and desorption kinetics was obtained via temperature programmed surface reaction techniques (refer to Figure 1e and Figure 1f).

Significance

The improved mixed oxide catalysts developed in this work show outstanding performance in ethanol oxidation and provide an excellent example for the sustainable synthesis of C2-building blocks which are necessary for the defossilization of chemical industry in the near future.

Acknowledgements

The authors gratefully acknowledge the funding by the Deutsche Forschungsgemeinschaft (DFG, German Research Foundation) - CRC 1487 – Project number 443703006. The authors also thank Marcus Rose, Kathrin Hofmann, Felix Reinauer and Christian Hess for their insights and their help with this study.

References

1. R. Melendez, B. Mátyás, S. Hena, D. A. Lowy, A. El Salous, *Renewable and Sustainable Energy Rev.* **2022**, *160*, 112260.
2. A. Dagle, A. D. Winkelman, K. K. Ramasamy, V. Lebarbier Dagle, R. S. Weber, *Ind. Eng. Chem. Res.* **2020**, *59*, 4843.
3. Oefner, F. Heck, M. Dürl, L. Schuhmacher, H. K. Siddiqui, U. I. Kramm, C. Hess, A. Möller, B. Albert, B. J. M. Etzold, *ChemCatChem* **2022**, *14*.
4. V. Raun, L. F. Lundegaard, J. Chevallier, P. Beato, C. C. Appel, K. Nielsen, M. Thorhauge, A. D. Jensen, M. Høj, *Catal. Sci. Technol.* **2018**, *8*, 4626.

Figure 1: (a): Conversion of ethanol (EtOH) and selectivity to acetaldehyde (AcH), acetic acid (AcOH), diethyl ether (DEE), ethene and CO_x at different reaction temperatures for an iron-molybdenum mixed oxide catalyst. (b): Ethanol conversion and selectivities to the different reaction products during a stability test at 240 °C. (c): Comparison of the selectivities to diethyl ether and ethene with the specific surface acidity of different iron-molybdenum mixed oxide catalysts. (d): Transient response of a catalyst during a pulse experiment used to study bulk oxygen diffusion. (e): Acetaldehyde signals during a temperature programmed desorption experiments performed at various heating rates. (f): Redhead-Plot of the desorption experiments used for the determination of the activation energy for acetaldehyde desorption.

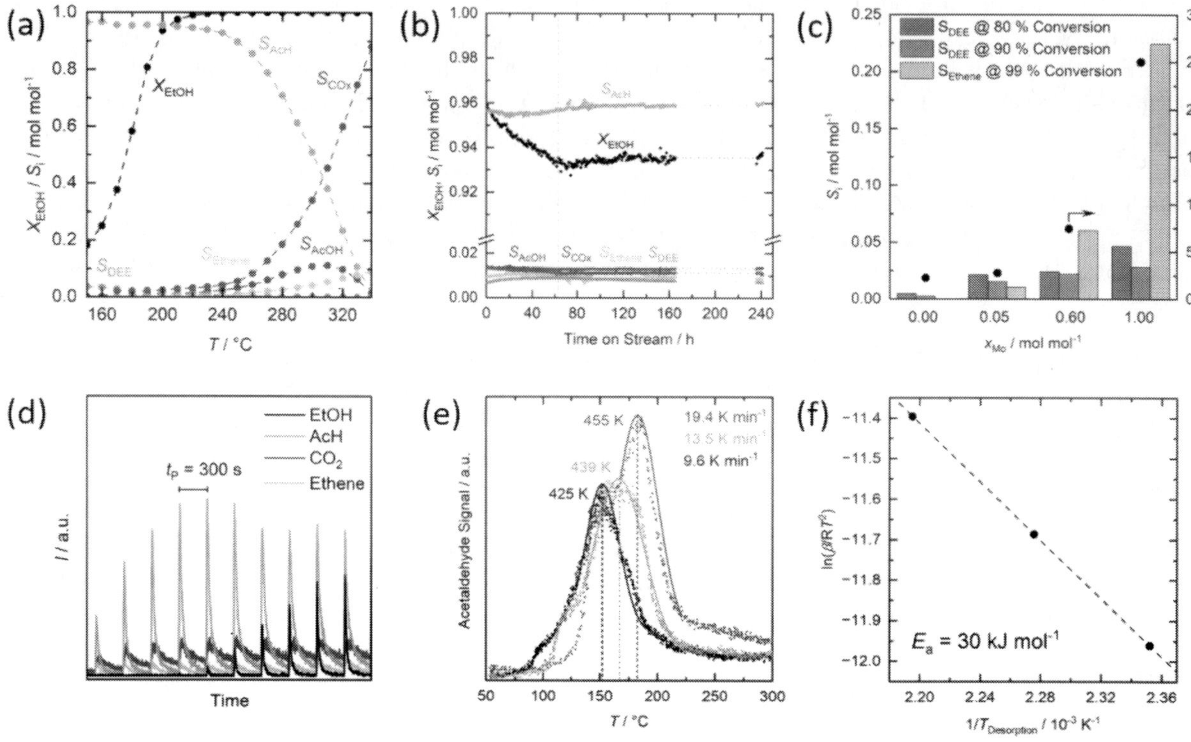
(a)
X_{EtOH} / S_i / mol mol^{-1}
T / °C
X_{EtOH}
S_{AcH}
S_{COx}
S_{AcOH}
S_{DEE}
S_{Ethene}
(b)
X_{EtOH}, S_i / mol mol^{-1}
Time on Stream / h
S_{AcH}
X_{EtOH}
S_{AcOH} S_{COx} S_{Ethene} S_{DEE}
(c)
S_{DEE} @ 80 % Conversion
S_{DEE} @ 90 % Conversion
S_{Ethene} @ 99 % Conversion
S_i / mol mol^{-1}
Acidic Centres / µmol m^{-2}
x_{Mo} / mol mol^{-1}
(d)
EtOH
AcH
CO_2
Ethene
t_P = 300 s
I / a.u.
Time
(e)
Acetaldehyde Signal / a.u.
455 K
439 K
425 K
19.4 K min^{-1}
13.5 K min^{-1}
9.6 K min^{-1}
T / °C
(f)
ln(β/RT^2)
1/$T_{Desorption}$ / 10^{-3} K^{-1}
E_a = 30 kJ mol^{-1}

329c Alcohol Assisted Synthesis of Methanol from Synthesis Gas

Guanfu Liu, Bruce Welt

Recently, the Consortium for Waste Circularity has advocated the use of "Regenerative Robust Gasification" to simplify waste collection and boost recycling rates. Robust Gasification refers to a type of gasifier that can accept everything in municipal solid waste (MSW) without the need for sorting. It promises to convert unsorted municipal waste, including multi-layer plastics, metallized flexible films, and many materials that are not conducive to traditional recycling, into synthesis gas (syngas). Since syngas is not easily transported, the conversion of syngas into a fungible, widely used chemical feedstock is proposed. As the backbone of C1 chemistry, methanol is an ideal target. Unlike the gas-phase methanol production widely used in industry, alcohol-assisted syngas to methanol process has different reaction mechanism, requires lower reaction temperature, therefore providing a new possible path.

329d Empowering Urea Electrosynthesis from Carbon Dioxide and Nitrate: Modulating *NO_2 Adsorption on Copper for Enhanced Performance

Bihao Hu, Lei Wang

In conventional electrochemical coupling reactions involving two distinct reactants, catalysts are typically required to exhibit high activity for both half-reactions while ensuring kinetic compatibility. However, in C-N coupling reactions where the valency of carbon or nitrogen remains constant in the reactants and products, achieving high production performance may not necessitate the catalyst to be equally active for both half-reactions. As the typical C-N bond-involved high-value chemical, urea commands significant market demand, which is a prominent fertilizer for agriculture. This study investigates this hypothesis within the framework of urea electrosynthesis, a process that directly converts carbon dioxide (CO_2) and nitrate (NO_3^-) into urea without altering the valence state of carbon.

Utilizing copper (Cu) as a catalyst, capable of catalyzing both CO_2 reduction and NO_3^- reduction, urea synthesis can occur within a potential region where merely NO_3^- reduction takes place, contrary to the expectation of kinetic matching. Through the implementation of density functional theory, we predict that the carbon intermediate involved in the C-N coupling process primarily originates from CO_2 rather than from CO_2 reduction. Moreover, we elucidate the pivotal approach to achieve high urea selectivity is regulating the competition of *NO_2 for coupling with CO_2 and H.

Furthermore, by incorporating a minute quantity of non-metallic atoms with high electronegativity, Cu-based catalysts are synthesized, significantly enhancing selectivity towards urea from below 20% to 80% with the highest yield rate of 6075 μg h^{-1} cm^{-2} for 100 h in a typical flow cell. The improvement predominantly stems from the mitigated adsorption of intermediates, making *NO_2 more predisposed to couple with CO_2 rather than undergoing hydrogenation. The reaction mechanism is further investigated through in-situ Fourier-transform infrared techniques. Scaling up the electrode into a 10 × 10 cm^2 flow cell, the overall current can escalate to 3.5 A while maintaining approximately 50% selectivity towards urea for a duration of 10 hours. This work provides an alternative strategy for designing catalysts for electrochemical coupling reactions.

329f Electrochemical Approach to Decarbonize the Ethylene Glycol Production

Rong Xia, Edward H. Sargent

Electrifying chemical productions is a potential approach to decarbonizing the chemical industry. Currently, 21% of the global greenhouse gas emissions come from the industry sector, while chemical manufacturing accounts for the largest share of industrial carbon emissions. The electrochemical processes, when powered by renewable electricity, have lower carbon footprints than conventional thermochemical routes. The rapid development of renewable energy provides opportunities in electrifying chemical production and decarbonizing chemical industry.

In this presentation, I would like to share our recent progress on electrifying ethylene glycol production with low carbon footprint.[1] Ethylene glycol is a $25B/year commodity chemical, mainly used as precursor of polyethylene terephthalate (PET) and antifreeze, responsible for the emission of 46 MMt CO_2/year. Its present-day production via thermocatalytic ethylene oxidation leads to a carbon intensity of ~ 1.2 CO_2eq per tonne of ethylene glycol. Here we report an electrochemical approach for ethylene glycol electrosynthesis with >90% selectivity towards ethylene glycol at a current density of 200 mA cm^{-2}. The reaction mechanism is investigated through surface-enhanced Raman spectroscopy, isotopic-labeling experiment, and density functional theory calculation.

References

1. **Xia, R.;** Wang, R.; Hasa, B.; Lee, A.; Liu, Y.; Ma, X.; Jiao, F., Electrosynthesis of Ethylene Glycol from C_1 Feedstocks in a Flow Electrolyzer. Nat Commun 2023, 14, 4570.

329g Hydrogen Direct Reduced Ironmaking: Technoeconomics of Titanomagnetite Ironsand Reduction in a High-Temperature Fluidised Bed

Matthew Watson, Cassidy van Vuuren, Bavinesh Maisuria, Sigit Prabowo, Diego Del Puerto, Raymond Longbottom, Brian Monaghan, Mark Reid, James T. Hinkley, Chris Bumby

Problem:

The manufacture of iron and steel is responsible for approximately 7% of global man-made carbon dioxide emissions. [1] The use of hydrogen as a reducing agent can substantially reduce carbon dioxide emissions from the ironmaking process when hydrogen is derived from renewable resources such as renewable electricity. Our work has shown that pellets of hydrogen direct reduced iron (H2-DRI) can be produced in a vertical shaft furnace, however the profitability is highly dependent on the cost of electricity. [2] The capital and operating costs associated with a H2-DRI process can be significantly reduced if the iron ore pelletising and sintering steps are eliminated.

Iron ore fines can be reduced relatively quickly when compared to pellet reduction in a shaft furnace. While experiments with iron ore fines in a fluidised bed using hydrogen at high temperatures (> 800 °C) have been carried out, several studies have reported sticking of the particles. [3-5] Agglomeration of the sticking particles caused the bed to defluidise, effectively shutting down the process.

Methods:

Here, we report results from the reduction of New Zealand (NZ) titanomagnetite (TTM) ironsands in a small-scale (100 g of ironsand) laboratory fluidised bed reactor at temperatures between 800 – 1000 °C using hydrogen flow rates up to 5 standard L/min (Figure 1). Hydrogen enters the vertical quartz tube and fluidises the ironsand after passing through a quartz frit inside the tube.

Iron oxide reduction in hydrogen is carried out through several stages as follows:

$$3\ Fe_2O_3 + H_2 \rightarrow 2\ Fe_3O_4 + H_2O \quad (1)$$

$$2\ Fe_3O_4 + 2\ H_2 \rightarrow 6\ FeO + 2\ H_2O \quad (2)$$

$$6\ FeO + 6\ H_2 \rightarrow 6\ Fe + 6\ H_2O \quad (3)$$

Overall Reaction: $3\ Fe_2O_3 + 9\ H_2 \rightarrow 6\ Fe + 9\ H_2O$ (4)

By considering the thermodynamic equilibrium between hydrogen, steam, and the oxidation state of iron of each step, the extent of ore reduction was tracked via a humidity sensor on the outlet of the reactor. Ironsand samples (~2 g) were extracted throughout the reduction run via a venturi sampling nozzle for analysis.

Extracted samples were characterized by powder X-ray diffraction [7] to determine which phases were present and the metallisation degree. Individual grains were analysed via scanning electron microscopy (SEM) and elemental energy-dispersive X-ray spectroscopy (EDS).

Results:

No sticking phenomena were observed, and a metallization degree of ~93% was achieved in 30 minutes using pure hydrogen at 950 °C. Sticking was inhibited by the formation of a protective Ti- and Al-rich outer oxide layer (Figure 2 and Figure 3) on the majority of particles, which prevented iron-iron contact.

To determine the economic viability of the process, we estimated the operating energy expenditures for the H2-DRI process carried out at 900 °C. As a first approximation we modelled each step of the process (Figure 4) through steady-state mass and energy balances. The original analysis was carried out assuming a counter-current vertical shaft reactor. However, the counter-current reactor can be approximated by a cascading series of fluidized bed reactors as described, for example, by POSCO's HyREX® process. [8]

The largest energy consumption in the process was associated with production of hydrogen (Figure 5) via electrolysis. Due to reaction equilibrium limitations, the reactor exit contains significant unreacted hydrogen (e.g. 46% hydrogen at 900°C). It is therefore essential to condense the steam and separate water from the unreacted hydrogen and recycle the hydrogen back to the reactor to continue to drive the reduction reactions and maintain economic viability.

Moreover, the cost of renewable electricity has a significant impact on the economic feasibility of the H2-DRI process. Figure 6 indicates that an input renewable electricity cost (at the H2-DRI production facility) of well below US$80/MWh is required if H2-DRI is to compete on a cost basis against existing carbothermic DRI processes. Due to the continuous nature and scale of a commercial H2-DRI process, it is important that electricity is sourced from firm (continuously operating) electricity generating facilities.

Implications:

The findings are important as increasing the operating temperature of the fluidised bed reactor beyond 800°C results in a faster reaction rate and higher gas utilisation, making the process more economically attractive. This has the potential to reduce the size of the reduction reactors and lower the capital costs of a H2-DRI facility in a final commercial-scale process. Moreover, obtaining long-term low-cost pricing contracts from firm renewable electricity generating assets (e.g. nuclear, geothermal, or hydroelectric) is required to ensure a profitable and therefore economically and environmentally sustainable process.

References:

1. https://ourworldindata.org/ghg-emissions-by-sector

2. Cassidy van Vuuren, Ao Zhang, James T. Hinkley, Chris W. Bumby, Matthew J. Watson, The potential for hydrogen ironmaking in New Zealand, *Cleaner Chemical Engineering*, Volume 4, 2022, 100075, ISSN 2772-7823, https://doi.org/10.1016/j.clce.2022.100075.

3. Zhan Du, Yu Ge, Fan Liu, Chuanlin Fan, Feng Pan, Effect of different modification methods on fluidized bed hydrogen reduction of cohesive iron ore fines, *Powder Technology*, Volume 400, 2022, 117226, ISSN 0032-5910, https://doi.org/10.1016/j.powtec.2022.117226.

4. Wolfinger T, Spreitzer D, Schenk J. Analysis of the Usability of Iron Ore Ultra-Fines for Hydrogen-Based Fluidized Bed Direct Reduction—A Review. *Materials*. 2022; 15(7):2687. https://doi.org/10.3390/ma15072687

5. Qiyan Xu, Haichuan Wang, Yuankun Fu, Jianjun Wang, Parameters Optimization, Sticking Mechanism and Kinetics Analysis of Fine Iron Ore in Fluidized-Reduction Process, *ISIJ International*, 2016, Volume 56, Issue 11, Pages 1929-1937, Released on J-STAGE November 17, 2016, Advance online publication September 13, 2016, Online ISSN 1347-5460, Print ISSN 0915-1559, https://doi.org/10.2355/isijinternational.ISIJINT-2016-280

6. Sigit W. Prabowo, Raymond J. Longbottom, Brian J. Monaghan, Diego del Puerto, Martin J. Ryan, Chris W. Bumby, Phase transformations during fluidized bed reduction of New Zealand titanomagnetite ironsand in hydrogen gas, *Powder Technology*, Volume 398, 2022, 117032, ISSN 0032-5910, https://doi.org/10.1016/j.powtec.2021.117032.

7. Prabowo, S.W., Longbottom, R.J., Monaghan, B.J. et al. Sticking-Free Reduction of Titanomagnetite

Ironsand in a Fluidized Bed Reactor. *Metall Mater Trans B*, 50, 1729–1744 (2019). https://doi.org/10.1007/s11663-019-01625-w

8. https://www.posco.co.kr/homepage/docs/eng7/jsp/hyrex/

Figure Captions:

Figure 1: Small-scale fluidised bed reactor.[6]

Figure 2: SEM images showing the microstructural evolution of grains of TTM ironsand during the reduction by 100% hydrogen at 950 °C: (a) raw ironsand, (b) 5 min, (c) 10 min, (d) 20 min, (e) 40 min, and (f) 60 min. [7]

Figure 3: EDS images showing elemental maps of a grain of TTM ironsand reduced in 100% hydrogen at 950 °C for 60 min. (a) is the BSE image, and (b) to (f) refer to the element noted in each image. [7]

Figure 4: Process flow diagram of hydrogen direct reduction of iron. [2]

Figure 5: Energy requirements for the hydrogen DRI process.[2]

Figure 6: Production cost for H2-DRI as a function of electricity cost. [2] Orange wedge represents the available processing margin for H2-DRI. The yellow region represents the additional available margin for billet steel from H2-DRI.

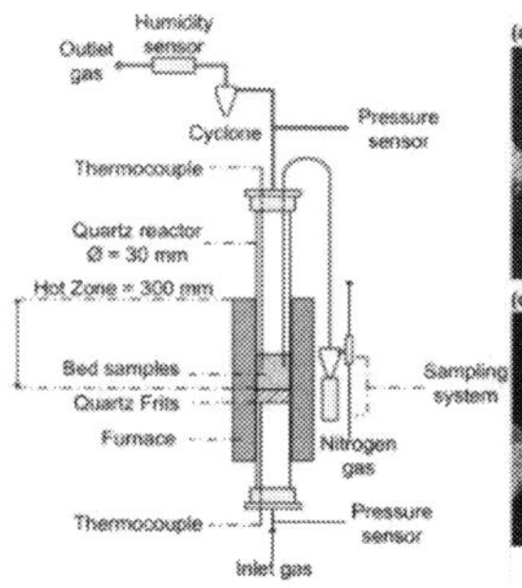

Figure 1

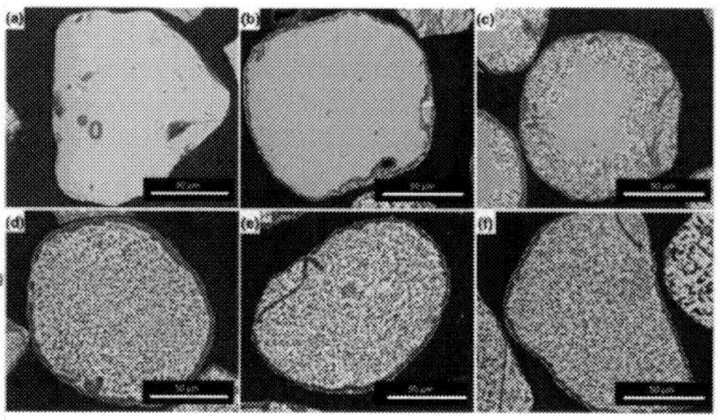

Figure 2

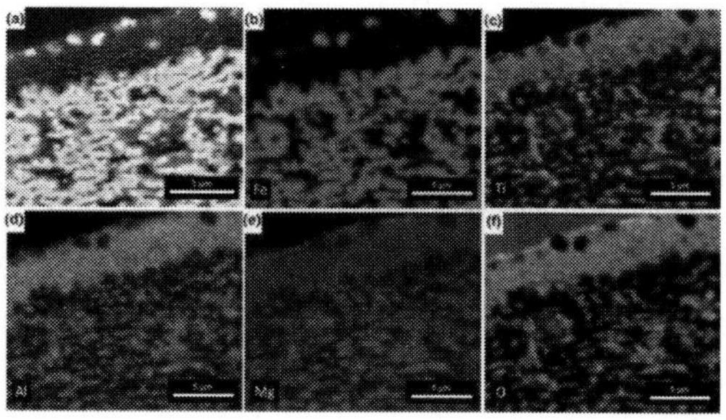

Figure 3

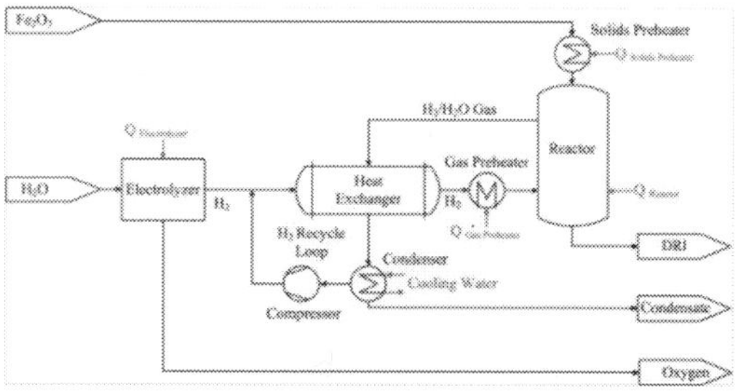

Figure 4

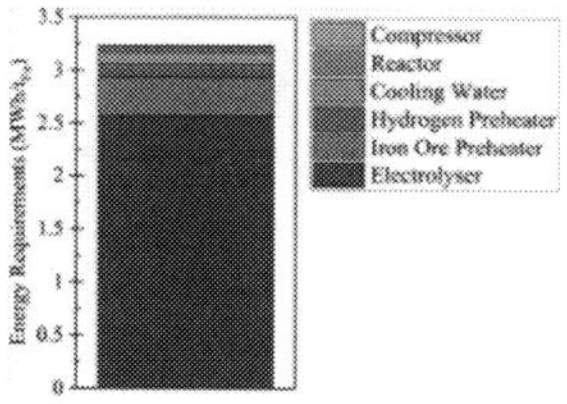

Figure 5

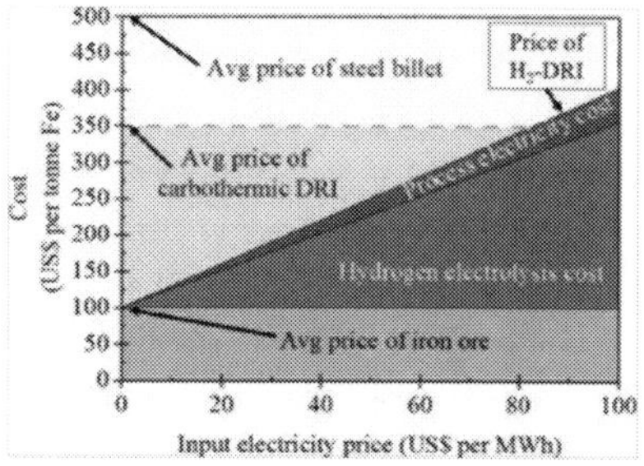

Figure 6

329h Concerted Methane Fixation at Ambient Temperature and Pressure Mediated By an Enzymatic and Fe-ZSM-5 Catalytic Couple

Jimin Kim, Daniel James Lundberg, Michael S. Strano

Anthropogenic methane emissions to the atmosphere pose a technological challenge for conversion into commodity chemicals due to their diffuse and dilute nature. However, the low temperature and concentration of these emission streams make them challenging to address by current routes of methane oxidation, which rely on high temperatures or pressures to drive methane oxidation rapidly and efficiently. Herein, we introduce a novel tandem catalytic system of alcohol oxidase with an iron-modified ZSM-5 that operates as a synthetic methanotrophic system capable of partially oxidizing methane at ambient temperatures and pressures, producing chemically useful intermediates for material synthesis. We show that the methane-to-formaldehyde selectivity can exceed 90% at room temperature. The generated formaldehyde intermediate can be rapidly incorporated into a growing urea polymer, with a material growth rate exceeding 5.0 mg g_{cat}^{-1} hr^{-1}, commensurate and even exceeding rates for many cultured methanotrophic bacteria systems. Strategies to enhance the stability of this tandem catalysis and technical analysis for future application scenarios will be discussed.

724a Non-Thermal Plasma-Assisted Hydrogenolysis of Waste Plastics to Light Hydrocarbons

Libo Yao, Steven S.C. Chaung, Zhenmeng Peng

While the wide application of plastic materials has significantly improved the convenience and quality of human life, the processing and recycling of waste plastics (WPs) has become an ever-growing challenge. Accompanied by the big demand for plastic products, more and more WPs have been generated. It is reported that more than 300 million tons of WPs were generated in 2015 alone. As of date, only less than 20% of these WPs can be recycled, with the rest ending up in landfills and incineration or being directly released to the environment. The inability to recycle WP has not only caused huge pressure on the environment, but also wasted tremendous amount of carbon resource. Chemical processes including pyrolysis and gasification have been actively investigated for converting WPs to value-added synthesis gas and carbonaceous materials. However, a high operating temperature (700–1300 K) is typically required to overcome the unfavorable thermodynamics and kinetics that makes these methods cost inhibitive. More recently, there have been a few studies on plastics hydrogenolysis. However, harsh reaction conditions of high temperature (600–700 K) and pressure (10–30 bar) and relatively long reaction time (> 6 h) were still needed, suggesting low hydrogenolysis kinetics.

Herein, we report a novel non-thermal plasma-enabled method which allows efficient hydrogenolysis of pure and post-consumer plastics under ambient temperature and atmospheric pressure condition. Non-thermal H_2 plasma provides a unique medium for generating reactive hydrogen species, primarily in form of ions and radicals, which can effectively break C-C bond in the polymeric structures and drive kinetically unfavorable reactions under mild reaction conditions. Light alkanes, predominately CH_4, C_2H_6 and C_3H_8 with >95% selectivity, were obtained under ambient condition (25 °C) as result of favorable thermodynamics and fast reaction kinetics in the non-thermal plasma-assisted hydrogenolysis of high-density polyethylene. Employment of this method to polystyrene generated liquid products and C_1–C_3 hydrocarbons as major gas products. Interestingly, ethylene, an important precursor in polymer industry, can be produced at a yield as high as 38 wt.%. Moreover, this non-thermal plasma-assisted hydrogenolysis method allows direct valorization of post-consumer plastic products without a need of pretreatments and shows minimal influences by the contaminants and impurities. These findings reveal a great promise of this new method for WP valorization application.

724b Upcycled (Polyvinyl Chloride) PVC Support Layer from Waste PVC Pipe for Thin Film Composite Nanofiltration Membranes

Atta Ur Razzaq, Milad Esfahani

Thin-film-composite (TFC) nanofiltration membranes represent the pinnacle of membrane technology in water treatment and desalination. These membranes typically consist of a polyamide (PA) selective layer and a membrane support layer, often constructed from commercially available materials like polyethersulfone (PES). However, an alternative approach exists that involves using different polymers, preferably upcycled waste polymers, as a viable support layer for TFC membranes. Successfully implementing the upcycling of waste plastics into high-value support membranes can significantly contribute to addressing plastic waste pollution. One of the main challenges associated with utilizing upcycled polymers as support layers is the potential impact on the polyamide selective layer of TFC membranes, subsequently affecting their performance in terms of permeability, rejection, and antifouling properties. This study demonstrates the successful fabrication of TFC membranes with a support layer crafted from upcycled waste PVC pipe. We conducted a comprehensive investigation into the effects of upcycled PVC on the structural and physicochemical properties of the polyamide layer. The ultrafiltration (UF) support membranes were fabricated from waste PVC pipe via non-solvent induced phase separation (NIPS). Subsequently, a polyamide layer was synthesized atop the Upcycled PVC membrane using interfacial polymerization (IP). The physicochemical properties and performance of the TFC membranes with upcycled PVC support layer were compared with membranes with research-grade (RG) PVC and commercial PES support layers. The results unveiled that the TFC membrane with upcycled PVC support layer exhibited higher water permeability (18.2 LMH/bar), in contrast to RG TFC (15.5 LMH/bar) and PES TFC membranes (13.7 LMH/bar). Furthermore, the salt rejection capabilities of the upcycled PVC TFC membrane were competitive and well within an acceptable range.

724c Upcycled Group IV Base Oil Lubricants Generated from Pyrolysis of Military Waste Polyethylene and Polypropylene Plastics

Sarvada Chipkar, Daniel Kulas, Topher Taylor, Emily Leverance, David Shonnard

Group IV base oil lubricants are one of the high value products produced by the petroleum industry which are used as synthetic oils for the purpose of lubrication over a wide range of temperatures due to their superior oxidative stability at elevated temperatures and lower pour points at extremely low temperatures. Traditionally, the petroleum industry uses synthetic chemical processes to generate ethylene through crude oil or natural gas cracking followed by polymerization to target a specific carbon range (typically C8-C12) that is further distilled to produce pure normal alpha-olefins (NAOs). These NAOs are then catalytically synthesized and hydrogenated to obtain lubricant grade molecular weight range (C30 trimers) called poly-alphaolefins (PAOs) that possess better tribological performance qualities such as viscosity and oxidation stability. However, one of the challenges for this process is overcoming contamination from sulfur and aromatic compounds that are naturally present in the crude petroleum feedstock. The fast pyrolysis process used in this study has the ability to produce the NAOs without sulfur or aromatic contamination due to the use of a purer feedstock in the form of waste plastic (high and low density polyethylene and polypropylene).

Fast pyrolysis of military waste polyethylene (PE) and polypropylene (PP) at elevated temperatures (450-550° C) thermally broke down the longer carbon chains into three products – wax, oil and gas - which were condensed into specific carbon molecular range using condensers operating at specific temperatures. In this study, the wax was collected at two temperatures 75° C and 250° C while the oil was collected at 25° C. Qualitative gas chromatography coupled with mass spectrometry showed that the wax and oil predominantly contained C10-C30 and C8-C17 molecular carbon ranges respectively in the form of paraffinic alkenes, predominantly, without sulfur or aromatics. These products were further fractionated in a distillation column to acquire C9-C12 normal alkene products which were catalytically coupled to form compounds with longer carbon branching typically possessing the C30 isomeric range mimicking the Group IV base lubricant oils. This presentation will report several measurements; kinematic viscosities at 40° C and 100° C, pour point, flash point, and thermal gravimetric analysis (TGA) were conducted to evaluate the thermal stability, differential scanning calorimetry (DSC) was used to estimate melting point, Fourier transform infrared spectroscopy (FTIR) & nuclear magnetic resonance (NMR) to predict the molecular arrangement and functional groups and, gel permeation chromatography for molecular weight estimation were used to evaluate the tribological properties of the obtained lubricant group IV base lubricant oil. This process is unique in its kind as it aims at generating a high-value synthetic lubricant from waste plastic without sulfur or aromatic contamination.

724d The Influence of Structure-Property Relationships in Polyethylene Terephthalate on Glycolytic Deconstruction

SriBala Gorugantu, Michele Valsecchi, Sabin Adhikari, Sanat K. Kumar, Linda J. Broadbelt

Efficient recycling of plastics requires a thorough understanding of complex deconstruction mechanisms intertwined with the structure-property relationships of these feedstocks for optimizing the conversion. Continuum models are limited to tracking the overall decomposition of lumped polymeric species and the temporal evolution of products and provide little information on the influence of the structure and morphological properties, such as crystallinity, on the depolymerization mechanisms. Tracking specific chain sequences and changes in the morphology is important when studying glycolytic deconstruction at temperatures ranging between the glass transition temperature and the melting point of semicrystalline polymers, like polyethylene terephthalate (PET). This study aims to develop a detailed Kinetic Monte Carlo (kMC) framework to gain insights into the structure-property relationships of PET and the impact on glycolytic deconstruction.

In this work, the initial chain length distribution of PET is determined by assuming a Schulz-Flory distribution function. The spherulitic morphology is then defined by the distributions of tails, ties, loops, and crystalline stems present in the PET chains, which are generated using random-walk simulations based on their probabilities of formation. KMC simulations of catalytic PET glycolysis at discrete time steps for reaction temperatures between 150-190 ^{o}C suggest a competition between the random, chain scission and repolymerization reactions at different reaction timescales. Mapping the reactions and distributions suggests that interlamellar amorphous domains like tie chains and loops undergo sequential random scission forming tails and disrupt the spherulitic morphology during the initial phase of glycolytic deconstruction. While the end chain scission of free amorphous domains and tails resulted in BHET monomer, it has been observed that the reactivity of crystalline stems governs the monomer yield at longer reaction times. Such frameworks are promising in developing predictive models for the solvent-based depolymerization of condensation polymers and gaining insights into the structure-property relationships.

724f One-Dimensional Model of a Two-Stage Molten-Metal Bubble Column Reactor for H_2 Production from Polypropylene Pyrolysis

Bang Thanh Le, Son Ich Ngo, Young-Il Lim, Young Jae Lee, Uen-Do Lee

A one-dimensional (1D) reaction model is presented for a two-stage molten metal bubble column reactor (2-stage MMBCR) used for hydrogen production from polypropylene (PP) pyrolysis. In the first stage reactor, solid PP decomposes into hydrogen and gas-phase hydrocarbons (C1-C4), which was modeled by an empirical equation. The second stage reactor for hydrocarbon pyrolysis in the MMBCR was represented by a one-dimensional plug flow reactor (PFR) model with non-catalytic and catalytic reactions. Catalytic reactions occurred at the gas-liquid interfacial area, while non-catalytic reactions took place within the bubble volume. The 2-stage MMBCR model was validated against experimental data over a wide range of operating temperatures. The H_2 yield was compared for 2-stage MMBCRs with Sn and Ga-Sn alloys. This study provides insights for designing the reactor size and optimizing the operating conditions of 2-stage MMBCRs to produce H_2 from waste plastics.

646a Mechanism-Based Lignin Bioprocess and Biomaterial Design

Zhihua Liu, Jinghao Li, Qiang Li, Jorge Arreola Vargas, LE Zhou, Furong Lin, Arthur Ragauskas, Bin Yang, Fujie Zhou, Susie Dai, Joshua Yuan

Lignin is one of the most abundant biopolymers on earth. Despite the abundance, lignin remains primarily a waste stream from paper, pulping, and biorefining industries. The utilization of lignin as a fungible product is fundamental to the sustainable development of our society. Nevertheless, efficient and cost-effective lignin processing and material heavily depend on the structure-functional relationship between lignin chemistry and bioprocess or biomaterial features. In the past decades, we have substantially advanced understanding of how lignin chemistry could impact bioprocessing efficiency and biomaterial quality. First, we have revealed that lignin fractionation into smaller molecular weight, monomer-containing, and hydroxyl group-enriched fractions will significantly improve lignin dissolution and bioprocessibility. Based on the fundamental understanding, we have substantially improved lignin conversion efficiency, titer, and economics to PHA and lipid, along with achieving efficient co-utilization of carbohydrates and lignin. Second, we discovered that the higher molecular weight, better uniformity, and more ether linkages could improve carbon material properties. The fundamental understanding has guided the lignin process and feedstock design to substantially improve lignin carbon fiber property and achieve co-improvement of feedstock stress tolerance. Third, we found that lignin with a more condensed structure improves the stability and size selection for nanoparticles. We subsequently designed biomass processing to derive the structure to produce stable lignin nanoparticles. Fourth, we found that lignin functional groups and molecular weight also impact the interaction with asphaltene in a way that lignin can be tailored to derive asphalt binder modifiers with higher PG grades. We have integrated the discovery with biorefinery design to derive lignin as an effective asphalt binder modifier. Overall, based on the fundamental understanding, we have developed various process and feedstock designs to transform lignin into a precursor for broader biomaterial applications. The fundamental understanding will guide the advancement of a broader spectrum of functional materials from lignin.

646b A Simultaneous Depolymerization and Hydrodeoxygenation Process to Produce Lignin-Based Jet Fuel in Continuous Flow Reactor

Adarsh Kumar, Abhishek Kumar, Daniel Santosa, Huamin Wang, Peng Zuo, Chongmin Wang, Rafal Gieleciak, Bin Yang

Economical production of lignin-based jet fuel (LJF) can improve the sustainability of sustainable aviation fuels (SAFs) as well as can reduce the overall greenhouse gas emissions. However, the challenge lies in converting technical lignin polymer from biorefinery directly to jet fuel in a continuous operation. In this work, we demonstrate a simultaneous depolymerization and hydrodeoxygenation (SDHDO) process to produce lignin-based jet fuelfrom the alkali corn stover lignin using engineered Ru-HY-60-MI catalyst in a continuous flow reactor at first time. The maximum carbon yield of LJF of 17.9 wt% was obtained and it comprised of 60.2 wt% monocycloalkanes, and 21.6 wt% polycycloalkanes. Catalyst characterization of Ru-HY-60-MI suggested there was no significant change in HY zeolite structure and its crystallinity after catalyst engineering. Catalyst characterizations performed post the SDHDO experiments indicate presence of carbon and K content in the catalyst. The K content presence in spent catalyst was due to K^+ ion was exchanged between lignin solution and HY-60 while carbon presence validated the SDHDO chemistry on the catalyst surface. Tier α fuel property testing indicates that LJF production using SDHDO chemistry can offer SAF with high compatibility, good sealing properties, low emissions, and high energy density for aircraft.

646d Multi-Stream Integrated Biorefinery for Sustainability and Economics

Jinghao Li, Jorge Arreola Vargas, Cheng Hu, Zhihua Liu, Phatchaya Piriyathanasak, Rongchun Shan, Yunyan Wang, Xianzhi Meng, Chengcheng Fei, Bruce McCarl, Bin Yang, Arthur Ragauskas, Susie Dai, Joshua Yuan

Abstract: Lignocellulosic biomass is an abundant natural resource that can be employed as a substitute for petrochemical products, a source of renewable energy, and a means to mitigate global climate change. Considering the substantial impact on sustainability, lignocellulosic biorefinery design has been thoroughly studied for the past decade[1-3]. Despite these efforts, the major challenge for biomass utilization remains to be the valorization of all cell wall components in a sustainable, economical, and integrated process. In particular, lignin valorization remains challenging due to its complex and heterogeneous structure. The fundamental structure-function relationship needs to be deciphered in the context of bioprocess design to address the challenges leading to this point. In order to address these challenges, we propose an innovative processing avenue form Multi-stream Integrated Biorefinery (MIBR) that is informed by how the lignin structure is changed by the fermentation process. The chemical structure further informs the process design based on the structure-function relationship. The novel biorefinery design allows the production of multiple value-added products, specifically, bioplastics and carbon fiber, using the same lignin, which substantially improved the economics and sustainability of biorefinery. Furthermore, we found that fermentation improves lignin structures to enhance carbon fiber performance, allowing synergistic process design based on chemical structure. Ultimately, we have further addressed the challenges of bioplastics quality by producing renewable carbon fiber reinforced bioplastics (RCFRP). Unlike conventional CFRP, the unique biorefinery design delivers a novel class of sustainable reinforced composite materials that can be biodegradable, and ensuring both sustainability and economic efficiency for the biorefinery industry.

646e Acidolysis of Lignin Model Compounds and Lignin Polymer Using CO_2-Gas Expanded Water

Anand Narani, Snehasish Mondal, Marcus Foston

This study explores acidolysis for lignocellulose fractionation, essential for aromatic compound production. Traditional acid treatments of lignin, especially under high temperatures and prolonged solvent exposure, often lead to low monomer yields and insoluble residues. Key factors identified include formaldehyde loss and the formation of unstable phenylethane aldehydes, which readily recondense in acidic conditions. These challenges, coupled with the environmental impact of acid use —requiring neutralization and resulting in significant waste—underscore the need for alternative approaches. In response to these issues, our research explores the utilization of CO_2-Gas Expanded water to mitigate the challenges associated with conventional acidolysis. This innovative approach leverages the reaction of carbon dioxide with water to produce carbonic acid, which then dissociates, increasing the concentration of hydronium ions in the solution. This acidification is instrumental in breaking ß-O-4 ether bonds in lignin or lignin model compounds, leading to the formation of phenylethane aldehydes. Notably, the solvation properties and acidity of CO_2-Gas Expanded water can be precisely controlled by adjusting the pressure and temperature, thus enhancing product separation and stability within the reaction medium. Our investigation primarily focuses on 2-phenoxy-1-phenylethanol as a model compound for lignin, aiming to demonstrate the efficacy and environmental advantages of this novel approach to lignin acidolysis.

646f Non-Isocyanate Synthesis of Lignin-Based Polyurethane Foam

Yunxuan Wang, Xianzhi Meng, Charles M. Cai, Eric A. Mintz, Chang Geun Yoo, Deepak Kumar, Obste Therasme, Shuai Li, Endong Wang, Arthur Ragauskas

Polyurethanes (PU) are the fifth most widely used polymeric materials in modern life, finding diverse applications in insulation, construction, furniture, and medical devices. However, conventionally, PU is produced from petroleum-based polyols and highly toxic diisocyanates. To achieve sustainable and green production of the PU foams, a non-isocyanate synthesis using lignin derived from biomass is developed. Lignin was first extracted via co-solvent enhanced fractionation (CELF) process, then further fractionated and characterized with NMR and GPC. The fractionated lignin was epoxidized and converted to carbonated lignin through a carbon-negative process. The non-isocyanate synthesis of PU was readily conducted by reacting carbonated lignin with diamines. A novel one-step lignin epoxidation process was developed, and the impact of lignin structure on epoxidation was investigated, yielding insights into the selection of starting materials, with the aim of achieving high-performance, eco-friendly, and cost-competitive biobased PU insulating foams.

646g Design Heterogenous Catalysts for Developing Lignin Conversion Processes Guided By Inverse Molecular Design Theory

Dequan Xiao

Designing sustainable heterogenous catalysts is the key for developing chemical processes to convert lignin into value-added chemicals such as biofuels. However, catalyst design remains a grand challenge in computation or experiment, due to the vast numbers of possible variables (e.g., chemical compositions and particle size of catalytic domains) in the chemical space of catalytic structure. In this work, we demonstrate an effective integrative approach of heterogenous catalyst design, guided by the inverse molecular design theory in the framework of tight-binding electronic structure theory [J. Phys. Chem. 123(46), **2019**, 10019]. The inverse molecular design (IMD) theory aims at searching for optimum catalytic properties using deterministic optimization techniques along the hypersurface of catalysis-structure, and then mapping out the catalytic structures at the optimum points, leading to a very efficient way to accelerate catalyst design. By this approach, we designed a novel CuFeZn alloy catalyst for converting lignin into value-added chemicals via hydrogenation. By catalyst characterizations and reaction analysis in experiments, the designed catalyst was verified with significantly improved activity compared to the reference Cu catalyst. This degraded lignin products can be used for biofuel production via hydrodeoxygenation. Our results indicated that the inverse molecular design approach is a promising artificial intelligent solution to accelerate the search of optimum catalysts for converting lignin to value-added chemicals. In addition, we will explore the principle of designing dehydrogenation catalysts for H_2 production from lignin derived chemicals guided by the inverse molecular design approach.

406a Developing Engineering and Societal Consensus on New Processes

William Green

In the next two or three decades most of today's large-scale chemical and fuel processes will need to be replaced by low-greenhouse alternatives. In most cases the new processes will be more expensive than the incumbent high-emitting processes. The only way this can happen if government policy favors introduction of the new process. But if government policy favors an expensive or ineffective new process, society might pay a high cost for little benefit. Clearly it would be beneficial if the future policies are based on an accurate understanding about which new processes are most beneficial to society; in an ideal world this accurate understanding would come from careful analyses by the engineering community. In this talk, I will consider the challenge this reality poses to the engineering community, and suggest how we might develop a better way of working together to provide the policy-makers and the general public the information needed in a timely manner.

406b Efficient Design and Optimization of Integrated Fuel Cell Systems for Rail Transportation

Maria Portarapillo, Augusto Bellucci Sessa, Almerinda Di Benedetto

Highlights

- Design of integrated systems for hydrogen production and fuel cells for rail transport
- Different hydrogen storage medium were analysed such as methanol, ethanol glycerol and ammonia.
- Alternative hydrogen storage results feasible in terms of weight and volume and safer.

1. Introduction

Due to the depletion of fossil fuels and the deterioration of environmental conditions, global attention is increasingly focussing on the use of renewable energy sources and the improvement of energy efficiency. The use of hydrogen-based technologies requires infrastructures for production, transport, storage and distribution on a small, medium and large scale. The main focus is on optimising technical performance. However, when developing hydrogen infrastructure, safety must be considered first and foremost. In this context, the proton exchange membrane fuel cell (PEMFC), which can efficiently convert hydrogen into electricity, is becoming increasingly important, especially in mobile systems (e.g. vehicles). The HT-PEMFC in particular is very promising thanks to the typical operating temperatures (120-200 °C) and the resulting CO tolerance (4-5 % at 200 °C). In addition, the improvement in reaction kinetics, catalysts tolerance, heat dissipation and water management due to the temperature increase also leads to higher energy efficiency of the HT-PEMFC. For the development of safe hydrogen storage, the in-situ production of hydrogen from 'safer' chemicals was proposed as an alternative.

The aim of this work is to investigate the feasibility of an integrated railway traction system in terms of volume and weight based on hydrogen storage in the form of liquid methanol, ethanol, glycerol or ammonia, hydrogen production by oxidative steam reforming (OSR) or decomposition and HT-PEMFC. The calculations were performed for configurations optimised for energy self-sufficiency. The results were compared with those obtained for a configuration with on-board hydrogen storage at 350 bar. For the comparison of the different fuels, a train was used that was to be equipped with two three-phase asynchronous electric motors driven by inverters, each with an output of 314 kW (total output 628 kW). It was assumed that the entire traction system is housed in a single carriage (traction unit), the centre carriage, as developed, for example, as part of the HydroFlex project in the UK.

2. Methods

The energy self-sufficiency assessment of the integrated system was carried out using AspenPlus®. This assessment took into account the presence of a first stage to preheat the reactants, the reaction phase and the cooling phase of the outlet stream to reduce the temperature to a level acceptable for HT-PEM. The heat emitted by the fuel cell was compared with the heat required in the heating phase. The heat emitted by the cooling phase was also taken into account in order to achieve the autothermal state of the system. In the preliminary design of the system, an electrical output of 628 kW was assumed, which is also supported by ad hoc dimensioned Li-ion batteries. In accordance with literature data, the efficiency of the HT-PEMFC was assumed to be 50 %. The volume and mass of the catalyst were calculated on the basis of kinetic data given in the literature ([1]–[3]). The calculated volume was multiplied by a factor of 10 to take into account the necessary catalyst dilution to ensure quasi-isothermal operating conditions. The reactor volume is assumed to be 2 to 3 times the volume of the already assessed catalyst bed, taking into account the void fraction, heat exchanger and auxiliary equipment. Some commercial data is taken into account for the volume of the HT-PEMFC. The volumes of the alcohol/water or ammonia tanks were determined taking into account the hydrogen production yields given in the literature ([1]–[3]).

3. Results and discussion

The electrical system, consisting of high-temperature PEM fuel cells and batteries, was dimensioned by simulating the Brescia-Edolo line (Italy) and evaluating the required electrical power, motor traction and travelling speed using conventional train kinematics equations [4]. The battery used has a capacity of 400 kWh, the power is based on the maximum instantaneous power and the fuel cells correspond to the electrical power of the motors. For the reference case, where hydrogen is stored at 350 bar, the total weight of the traction system is 20 tonnes, of which about 50% is the weight of the Type IV containers used for gas storage at this high pressure and 437 kg is the weight of the hydrogen alone. In the case of methanol and ethanol, the total weight of the traction system is less than in the reference case and amounts to 14 and 17 tonnes respectively. Both alcohols are stored under liquid conditions at room temperature and pressure and with water under stoichiometric conditions, which ensures much safer conditions than storing H_2 under high pressure.

In the case of glycerol, the optimum operating temperature of the reactor is 600 °C. Under these conditions, a system weight of around 22 tonnes was determined. In the case of ammonia, liquid storage at room temperature and 9 bar as well as a reactor operating temperature of 300 °C are required. Under these conditions, the system has a weight of 14 tonnes. All volumes are compatible with the size of a railway wagon and the weights are below those permitted on railway lines such as the Brescia-Edolo line. In the chemical storage of H_2, the influence of by-products such as N_2, CO, CO_2 and CH_4 produced during reforming or cracking should be considered. In fact, the presence of CO_2 in the reformate can lead to a drop in stack voltage, but the presence of CO_2 in the reformate can lead to a drop in stack voltage, but it has a limited effect on the electrical efficiency of the stack [5]. In general, hydrogen-rich reformate with a purity of >50 % can be utilised directly by HT-PEMFCs without further purification processes. The reduction in hydrogen utilisation and PEM efficiency was taken into account, and the storage volumes were also updated and discussed.

4. Conclusions

The sizing of the proposed integrated systems with high-temperature PEM fuel cells, batteries and H_2 production step in the Brescia-Edolo route simulation has provided valuable insights into the potential application of different and safe hydrogen storage solutions. The comparison of hydrogen storage methods, including storage at high pressure, under liquid conditions and chemical processes such as reforming or cracking, revealed clear trade-offs in terms of weight, operating temperatures and efficiency. These results highlight the complexity of selecting an optimal hydrogen storage solution, with factors such as weight, operating conditions and by-product management playing a crucial role. The results contribute to ongoing efforts to develop sustainable and efficient electrical systems for train traction and have potential implications for future developments in this area.

References

- J. Kim, T. S. Nguyen, J. Mahmood, and C. T. Yavuz, Chem. Eng. J., 463 (2023) 142474
- Bac, S. Keskin, and A. K. Avci, Int. J. Hydrogen Energy, 45 (2020) 34888–34917

- Selva Roselin and H. W. Chiu, J. Saudi Chem. Soc., 22 (2018) 692–704
- Joshua Peli, "APPLICAZIONE DELLE FUEL CELL PER L'ALIMENTAZIONE DI VEICOLI FERROVIARI," Politecnico Milano, 2018.
- Guan, P. Alvfors, and G. Lindbergh, Appl. Energy, 130 (2014) 685–691

Acknowledgment

This study was carried out within the MOST – Sustainable Mobility National Research Center and received funding from the European Union Next-GenerationEU (PIANO NAZIONALE DI RIPRESA E RESILIENZA (PNRR) – MISSIONE 4 COMPONENTE 2, INVESTIMENTO 1.4 – D.D. 1033 17/06/2022, CN00000023). This manuscript reflects only the authors' views and opinions, neither the European Union nor the European Commission can be considered responsible for them.

406c Innovative Design and Performance Assessment of a Novel Modular Reactor for One-Step Liquid Fuel Production from Stranded Natural Gas

Tanay Jawdekar, Sudeshna Gun, Congwen Lu, Anuj Joshi, Sonu Kumar, Joel Paulson, Liang-Shih Fan

With increasing global energy demands, the market for fossil-based fuels is projected to grow substantially over the next 30 years. Oil and gas, being the major fossil-based energy sources for the transportation and industrial sector, have demonstrated an increased crude extraction by ~120% in the past decade. However, the crude extraction process faces a challenge in tackling the associated natural gas that is dissolved in the crude oil or prevails in the free gas caps of oil reservoirs. Considering the risks and costs involved in natural gas transportation, the quantity of associated gas is very small to substantiate the construction of a pipeline network. Although this associated natural gas is the by-product of the oil extraction process, the absence of a viable solution for utilization makes it 'stranded natural gas (SNG).' Approximately 15 tons/day of SNG is flared from one oil well, contributing to greenhouse gases. The uneconomical nature of pipeline installation to transport SNG and the necessity to find a robust solution to tackle SNG opens an avenue for developing modular technologies that can convert SNG to value-added products while reducing CO_2 emissions. Liquid fuels (LF) are high-density energy carriers that are easy to transport and can support the existing infrastructure for energy consumption. 15 tons/day of SNG that is flared has the potential to produce ~2.5 tons/day of LF and generate ~115 kW of electricity simultaneously. This work proposes a single reactor modular unit that converts SNG to value-added LF. The novel reactor configuration consists of a multi-tubular packed bed setup divided into three intermediate sections: a mixed reforming section, a section for heat exchange, and a Fischer-Tropsch (FT) section. Mixed reforming of SNG uses CO_2 and steam to produce high-quality syngas (H_2:CO = 1.7), which in turn can be used to produce liquid fuels in the FT section of the reactor. The heat exchange section assists in increasing the efficiency of the process by facilitating heat integration within the process streams. This process provides inherent CO_2 utilization and converts pressurized SNG to easy-to-transport liquid fuels that can be processed to manufacture motor spirit, diesel, jet engine fuels, and feedstocks for the petrochemical industry. System-level thermodynamic evaluation was conducted using ASPEN Plus software for the production of 2.5 tons/day of liquid fuel by taking into consideration multiple plausible cases of reactor configuration and heat integration. Detailed economic calculations performed based on the tubular reactor design for all configurations indicate that the production of liquid fuels using SNG can prove to be a profitable venture. The design of the multi-tubular reactor and the economics associated with it are simultaneously optimized using the Bayesian optimization technique to propose a portable, modular reactor requiring a minimum number of units to process the SNG from an oil well. The robustness of the best-case scenario is then validated for fluctuating SNG compositions, flow rates, and temperatures. This work establishes the design feasibility of a novel, economically viable, modular technology for upgrading SNG to value-added liquid fuels while reducing greenhouse gas emissions.

406d Design and Techno-Economics of Hydrogen Transportation in Saudi Arabia

Fatimah Alhassan, Umer Zahid

As global energy demand increases and climate change concerns worsen, hydrogen emerges as a viable alternative energy carrier due to its versatility and potential for zero-emission applications. The geographic location of Saudi Arabia makes it an ideal candidate to unlock the potential of hydrogen energy. Nevertheless, the development of cost-effective storage solutions capable of accommodating large quantities of hydrogen is a serious challenge. Safety and efficiency factors add to the complexity of hydrogen production and utilization. This research investigates the issues of long-distance pipeline transportation of hydrogen throughout Saudi Arabia, which serves a variety of commercial and industrial purposes. It evaluates the techno-economic feasibility of various hydrogen transportation scenarios by tracing routes from production locations to consumer terminals. Three principal hydrogen production facilities are under consideration: green hydrogen production in Neom via water electrolysis, blue hydrogen generation in Jubail via natural gas-based steam methane reforming, and blue hydrogen production in Jazan via heavy oil gasification. The study examines the transportation of high-purity hydrogen from the production site in a compressed state (at 210 bar) across a distance of 500 km from its initial location. The results indicate that manufacturing blue hydrogen by steam methane reforming in Jubail has the lowest production cost, whereas green hydrogen production in Neom has the lowest carbon emissions. Blue hydrogen generation in Jazan, on the other hand, produces the most CO2, although it is less expensive than green hydrogen production. The cost of hydrogen transportation varies according to parameters such as hydrogen source, transit distance, ambient conditions, geography, as well as pressure and the phase at the delivery site. This study gives vital insights into the technological, economical, and environmental elements that are critical for Saudi Arabia's energy industry. Such insightful findings are crucial to achieving the goals for the 2030 country's Vision for a hydrogen-based economy.

406e Identification of the Steel Decarbonization Options for Different Regions

Chao-Lun Mai, Haoxiang Lai, Bosong Lin, Guiyan Zang

Iron and steel manufacturing stands as a primary contributor to global CO2 emissions and ranks as the second-largest energy consumer among heavy industries. In the past decade alone, this sector has accounted for over 7% of global greenhouse gas emissions. Consequently, there exists an urgent imperative to identify practical pathways for substantial decarbonization. This research aims to analyze selected countries among the top eight steel producers –the U.S., Brazil, Germany, Turkey, China, India, Japan, and South Korea – to delineate practical decarbonization strategies.

Our analysis evaluated the impact of technology replacement, fuel switching, and carbon capture, utilization, and storage (CCUS) on energy demand, costs, and emissions in crude steel production. We employed two fundamental methodologies: Techno-economic Analysis (TEA) and Life Cycle Analysis (LCA). Technology replacement explored alternatives such as state-of-the-art blast furnace-basic oxygen furnace (BF-BOF-SOA), direct reduced iron with electric arc furnace (DRI-EAF and Hybrid-DRI-BOF) to replace the current blast furnace-basic oxygen furnace (BF-BOF) reliant on iron ores, as well as state-of-the-art electric arc furnace (EAF-SOA) to replace the current electric arc furnace (EAF) reliant on recycled steels. Fuel switching involved renewable electricity, renewable natural gas, and hydrogen. CCUS options focused on mono-ethanolamine (MEA). By factoring in regional-specific considerations such as materials and fuel inputs, electric power generation and network, green resource limitations, accessibility to new technologies, and policy constraints, we aimed to provide a recommended steel decarbonization pathway at the sector level with an optimized technology mix spanning from 2025 to 2050. We have also examined various scenarios under business as usual, carbon pricing, and incentives. Through this comprehensive analysis, our research endeavors to illuminate the most pragmatic and region-specific strategies for achieving deep decarbonization in the steel industry, thus offering a critical contribution to addressing the urgent global need for sustainable steel production.

406f Desing and Rating of a Thermically Integrated Distillation System with Low Stillage Generation

Gianpaolo Barci, Paulo S. Barci, Murilo Innocentini

Although ethanol is widely recognized as a renewable, sustainable, and secure alternative energy source in face of uncertaintites surrounding fossil fuels and imminent energy crisis, the substantial amount of ethanol produced in Brazil has resulted in a significant increase of stillage generated during the process. This poses a serious environmental issue due to its high pollution potential. Stillage is a rich mineral source and can be utilized as fertilizer in sugarcane fields, however, its limited storage lifespan at the mill and dosages restrictions led by environmental standards, transporting stillage across the field became costly beyond a certain distance from the distillery leding to a problem that industry urges to solve. Usually, stillage is produced in a quantity up to 12 times larger than ethanol, making it the most prevalent by-product of the process. To concentrate stillage, conventional mills requires specialized equipment which can account for up to 60% of the total capital expenditure of a distillery and often requires additional steam usage, which is not practical for the majority of the industry. The proposed solution employed heat integration techniques between multiple-effect stripping columns to effectively decrease the volume of stillage during the distillation process, rahter than after as emplyed today. Simulations using Aspen Plus indicated that the stillage/ethanol ratio dropped from 12:1 to 5:1 in volume without requiring extra steam or equipment. A subsequent rate-based simulation considering physical and hydraulic parameters was conducted to determine if these findings could be effectively translated in actual hardware with positive outcomes. These results regarding mass, energy and hydraulic balances are unprecedented and potentially groundbreaking in process engineering, leading to an innovative approach if the system could be built.

406g Techno-Economic Optimization of a Solvent Absorption Process for CO_2 Capture with 3d-Printed Intensified Packing

Stephen Summits, Debangsu Bhattacharyya, Grigorios Panagakos, Benjamin P. Omell, Michael S. Matuszewski

A significant challenge in implementing post-combustion CO_2 capture technologies for commercial use is the high cost. For a typical MEA process, costs can be upwards of 80 USD/ton[1]. A majority of the cost for these solvent-based absorption processes can be attributed to the operation of the reboiler in the solvent regeneration process, which typically requires 3.7 MJ of steam for every kilogram of CO_2 due to a high heat of absorption, upwards of 100 kJ/mol CO_2[2]. Although much work on developing novel solvents that have properties of higher capacity and lower heat of absorption has been done, considerable energy is still needed. Therefore, these systems should be operated as close as to their maximum thermodynamic efficiency for minimizing the energy requirements.

Heat generated from CO_2 absorption increases the temperature of the process, which in turn reduces the driving force for mass transfer[2]. To reduce the temperature bulge, intercoolers between beds of the absorption tower[3,4] are often used to cool the solvent. Intercoolers, however, are incapable of maintaining an optimal temperature profile since they are implemented at discrete locations between the beds. A potentially better solution is to utilize 3d-printed intensified packings with internal channels for flow of a cooling medium, thus allowing for simultaneous heat and mass transfer within the process[5]. Such devices can help to operate the towers close to their maximum thermodynamic efficiency. In this work, we also investigate the application of the intensified device in the stripper. Instead of providing the entire heat at the reboiler, which creates a considerably non-uniform temperature profile across the tower, the intensified packing can help to operate the stripper at the optimal temperature profile.

Rigorous properties models are used in this work. Modified e-NRTL model is used as the thermodynamic model[2]. Validated models for hydraulics, transport properties, heat transfer, and interfacial area are also implemented within the tower models. Optimal design and operation of the towers and the balance of the plant for the solvent-based CO_2 capture system are accomplished by solving a mixed integer non-linear programming problem by considering a flowsheet level model of the capture process[6]. The model is developed using the IDEAS modeling framework and considers a costing model to accurately evaluate the tradeoff between the capital and operating costs of the process[7]. The results show that the intensified packing when used in both absorber and stripper can lead to significant reductions in capture cost.

Acknowledgement

The authors graciously acknowledge funding from the U.S. Department of Energy, Office of Fossil Energy and Carbon Management, through the Carbon Capture Program.

Disclaimer

This project was funded by the Department of Energy, National Energy Technology Laboratory an agency of the United States Government, through a support contract. Neither the United States Government nor any agency thereof, nor any of its employees, nor the support contractor, nor any of

References

(1) Raynal, L.; Bouillon, P.-A.; Gomez, A.; Broutin, P. From MEA to Demixing Solvents and Future Steps, a Roadmap for Lowering the Cost of Post-Combustion Carbon Capture. *Chemical Engineering Journal* **2011**, *171* (3), 742–752. https://doi.org/10.1016/j.cej.2011.01.008.

(2) Morgan, J. C.; Chinen, A. S.; Omell, B.; Bhattacharyya, D.; Tong, C.; Miller, D. C. Thermodynamic Modeling and Uncertainty Quantification of CO2-Loaded Aqueous MEA Solutions. *Chemical Engineering Science* **2017**, *168*, 309–324. https://doi.org/10.1016/J.CES.2017.04.049.

(3) Chang, H.; Shih, C. M. Simulation and Optimization for Power Plant Flue Gas CO2 Absorption-Stripping Systems. *Separation Science and Technology* **2007**, *40* (4), 877–909. https://doi.org/10.1081/SS-200048014.

(4) Biliyok, C.; Lawal, A.; Wang, M.; Seibert, F. Dynamic Modelling, Validation and Analysis of Post-Combustion Chemical Absorption CO2 Capture Plant. *International Journal of Greenhouse Gas Control* **2012**, *9*, 428–445. https://doi.org/10.1016/J.IJGGC.2012.05.001.

(5) Bolton, S.; Kasturi, A.; Palko, S.; Lai, C.; Love, L.; Parks, J.; Xin, S.; Tsouris, C. 3D Printed Structures for Optimized Carbon Capture Technology in Packed Bed Columns. *Separation Science and Technology (Philadelphia)* **2019**, *54* (13), 2047–2058. https://doi.org/10.1080/01496395.2019.1622566.

(6) Akula, P.; Eslick, J.; Bhattacharyya, D.; Miller, D. C. Model Development, Validation, and Optimization of an MEA-Based Post-Combustion CO2Capture Process under Part-Load and Variable Capture Operations. *Industrial and Engineering Chemistry Research* **2021**, *60* (14), 5176–5193. https://doi.org/10.1021/acs.iecr.0c05035.

(7) Miller, D. C.; Siirola, J. D.; Agarwal, D.; Burgard, A. P.; Lee, A.; Eslick, J. C.; Nicholson, B.; Laird, C.; Biegler, L. T.; Bhattacharyya, D.; Sahinidis, N. V.; Grossmann, I. E.; Gounaris, C. E.; Gunter, D. Next Generation Multi-Scale Process Systems Engineering Framework. In *Computer Aided Chemical Engineering*; Eden, M. R., Ierapetritou, M. G., Towler, G. P., Eds.; 13 International Symposium on Process Systems Engineering (PSE 2018); Elsevier, 2018; Vol. 44, pp 2209–2214. https://doi.org/10.1016/B978-0-444-64241-7.50363-3.

406i Energy Storage Strategies for Integrating Chemical Plants with on-Site Renewable Electricity Generation

Shuaikang Du, Rakesh Agrawal

Electrification and decarbonization of the chemical industry are the keys to achieving carbon neutrality for human society, which necessitates the transition from the fossil-based chemical industry to the renewable-based chemical industry. To facilitate this transition, it is crucial to integrate renewable energy such as solar and wind energies into chemical processes. However, the intermittent nature of renewable energy requires storage equipment to provide a consistent and stable power supply.

Here, we focus on using on-site solar and wind power plants and energy storage equipment to deal with intermittency in renewable energy for energy-intensive decarbonized liquid fuel production from shale gas. Meanwhile, we analyze different strategies of energy storage, based on local sunlight intensity data, and identify the minimum storage requirement to operate the decarbonized process.

To reduce the storage requirement for the decarbonized process, we propose an energy storage system and build a linear programming model. Different factors have been investigated, including renewable energy curtailment, natural gas turbine, chemical turn-off, power grid connection, renewable mix, and electrolyzer capacity.

Our results provide useful insights into the strategies needed for energy storage volume and associated cost reductions in the context of decarbonized chemical plants. The methodology adopted in this study is general and can be applied to other decarbonized processes during the energy transition.

406j An Integrated Approach for the Recovery of Marine Chemicals: Technoeconomic and Sustainability Assessment

Barnabas Gao, Parul Sahu, Kirti Yenkie

Natural brines: sea brine, inland brine and lake brine are a major source for several mineral commodities and chemicals including calcium, sodium, potassium and magnesium salts and bromine. These mineral salts of sodium, calcium, potassium, and magnesium of sea origin are called marine chemicals and have high economic and strategic importance. Land-based minerals are being exploited to recover magnesium and potassium salts. Natural brines and halite rocks have been a conventional source for NaCl production using solar evaporation which accounts for 47% of global common salt production, multi effect evaporation and wet or dry mining technologies. This mineral processing results in a significant amount of by products, which remain unutilized in most cases. The utilization of saline effluent for resource recovery have attracted interest in recent times [1]. Existing brine mining technologies have been listed in literature and these include evaporation with sequential crystallization, membrane-based separation, chemical precipitation, etc. [2].

Sea bittern is a concentrated brine generated from sea salt production using solar evaporation. Several research has been conducted in regards to developing processes to recover some salts from bittern notably sodium, magnesium and potassium salts [2], [3]. The extraction of these valuable compounds is an emerging area of interest with significant potential for resource recovery and environmental sustainability. The reported processes in literature for the recovery of these salts from brine are treated as separate problems and as such there is no generic framework for this resource recovery. This study presents a comprehensive analysis into the development of a framework for optimal separation pathways for the recovery of valuable marine chemicals from brine employing a combination of process optimization and superstructure optimization techniques [4], [5]. A mixed-integer nonlinear programming framework is utilized to explore a diverse set of separation technologies, including evaporation, Centrifugation, Cooling Crystallization etc. The superstructure optimization approach enables the evaluation of various process configurations and the identification of the most cost-effective and environmentally friendly solutions for the recovery of these target chemicals. Techno-economic analysis (TEA) is conducted to assess the economic feasibility of the proposed separation pathways, considering factors such as capital investment, operational costs, and potential revenue streams. Life cycle assessment (LCA) is integrated into the recovery framework to evaluate the environmental metrics of each separation technology, including global warming potentials, human health impacts, ecosystem quality, and resource utilization.

The result of this study highlights the potential of utilizing an integrated process techniques to develop efficient, cost effective and sustainable processes for the recovery of valuable compounds and chemicals from saline systems. The integration of TEA and LCA into the developed framework provides a holistic understanding of the economic and environmental implications of different separation technologies, guiding the selection of the most feasible pathway for the separation and recovery of these marine chemicals.

References

[1] A. Panagopoulos, "Water-energy nexus: desalination technologies and renewable energy sources," *Environ. Sci. Pollut. Res.*, vol. 28, no. 17, pp. 21009–21022, May 2021, doi: 10.1007/s11356-021-13332-8.

[2] P. Sahu, B. Gao, S. Bhatti, G. Capellades, and K. M. Yenkie, "Process Design Framework for Inorganic Salt Recovery Using Antisolvent Crystallization (ASC)," *ACS Sustain. Chem. Eng.*, Dec. 2023, doi: 10.1021/acssuschemeng.3c05243.

[3] P. Sahu, "A comprehensive review of saline effluent disposal and treatment: conventional practices, emerging technologies, and future potential," *Water Reuse*, vol. 11, no. 1, pp. 33–65, Oct. 2020, doi: 10.2166/wrd.2020.065.

[4] K. M. Yenkie, W. Wu, R. L. Clark, B. F. Pfleger, T. W. Root, and C. T. Maravelias, "A roadmap for the synthesis of separation networks for the recovery of bio-based chemicals: Matching biological and process feasibility," *Biotechnol. Adv.*, vol. 34, no. 8, pp. 1362–1383, Dec. 2016, doi: 10.1016/j.biotechadv.2016.10.003.

[5] J. D. Chea, A. L. Lehr, J. P. Stengel, M. J. Savelski, C. S. Slater, and K. M. Yenkie, "Evaluation of Solvent Recovery Options for Economic Feasibility through a Superstructure-Based Optimization Framework," *Ind. Eng. Chem. Res.*, vol. 59, no. 13, pp. 5931–5944, Apr. 2020, doi: 10.1021/acs.iecr.9b06725.

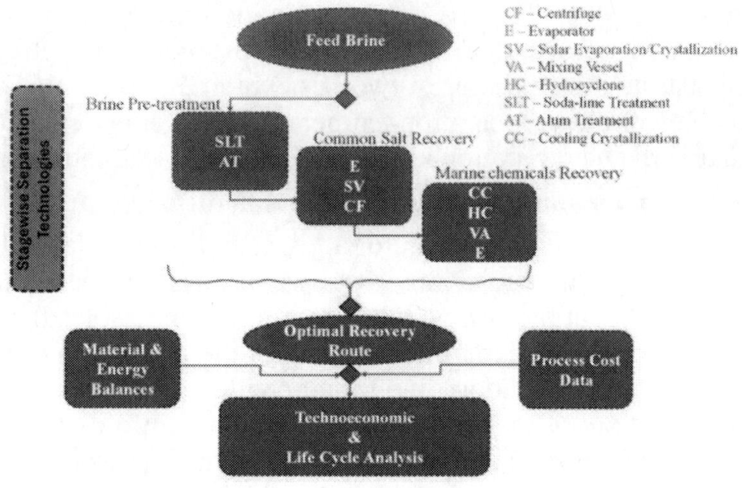

Figure 1: Framework for the Recovery of Marine Chemicals

537a Understanding the Mixing Behavior and Phase Separation Kinetics of Thermally Responsive Ionic Liquids in Water

Ahmed Mahfouz, Jordan Kocher, Akanksha Menon

Thermally responsive ionic liquids (ILs) exhibit a liquid-liquid phase separation in aqueous solutions when heated above a lower critical solution temperature (LCST). The phase separation results in the formation of a water-rich phase (WR) and an ionic liquid-rich phase (ILR). The chemical potential difference between the two phases can be leveraged for a wide range of applications from desalination to air conditioning and solvent extraction. With a focus on desalination as the application, desirable properties include a high osmotic strength (low water activity), tunable phase separation temperature, and fast phase separation kinetics. However, these properties are often inversely related as more hydrophilic ILs exhibit a higher osmotic strength at the cost of a higher phase separation temperature. To this end, we demonstrate that mixtures of two ILs exhibit a synergistic interaction, resulting in a higher osmolality and lower phase separation temperature than either constituent. We report the osmolality, phase diagram, and separation kinetics of N_{4444}Sal (tetrabutylammonium-salicylate) and P_{4444}TFA (tetrabutylphosphonium-2,4-trifluoroacetate) in different mixing ratios. This enables the selection of the ideal IL mixture composition for a given application depending on the imposed temperature and osmotic strength constraints. Next, we discuss the micro- to macro-scale phase separation behavior of the ILs at the critical temperature, which enables the development of a Stokes-Einstein-based kinetic model using experimental data for the nucleation behavior of the ILR phase (from microscopy). The phase separation time for a given IL mixture is measured using a light transmission technique that provides both spatial and temporal information. Results indicate that the time required for phase separation of a given IL mixture is analogous to the behavior of emulsions and obeys Stokes' law for a given phase boundary height and ILR phase nucleate size distribution. Overall, this work establishes an improved understanding of thermally responsive ILs and mixtures thereof for applications in energy and sustainability.

537b Spectroscopic Investigation of Silica Nanopore Confinement Effects on a Halometallate Ionic Liquid

Stephen Rankin, Andrew Drake, Barbara L. Knutson

Halometallate ionic liquids (ILs) combine the tunable solvent characteristics of organic salts that melt near ambient conditions (ILs) with catalytic and electrochemical activity of the metal centers. When they contain earth-abundant metals (iron, zinc, and aluminum), they represent potential green catalysts for a wide variety of Lewis acid-catalyzed reactions. However, to minimize the volume of solvent needed, to promote reusability, and to overcome transport limitations, halometallate ILs may need to be used as supported systems confined in nanoporous materials such as silicon dioxide (silica).

Here, the effects of confinement of ionic liquid (IL) 1-butyl-3-methylimidazolium tetrachloroferrate [BMIM][$FeCl_4$] within 3.5 nm pores of mesoporous silica thin films are investigated using x-ray photoelectron spectroscopy (XPS). Mesoporous silica thin films with accessible pores oriented orthogonally with respect to the substrate were synthesized by evaporation induced self-assembly using cetyltrimethylammonium bromide (CTAB) as the template, and characterized by grazing-incidence small-angle x-ray scattering. The IL was physically confined within the pores and XPS used to probe the interface between the IL and the silica support. The IL was also mixed with pyridine and supported onto 8 nm mesoporous silica microparticles to evaluate the acidity using FTIR. FTIR showed that compared to [BMIM] chloride, the addition of iron gave the IL Lewis and Brønsted acidity properties that were maintained when confined within the pores. Comparing [BMIM][Cl] to [BMIM][$FeCl_4$] using XPS showed that the charge of anion was more delocalized with the addition of iron and was less able to accept hydrogen bonds. The confined [BMIM][$FeCl_4$] interacted with the pore wall in a different way than [BMIM][Cl]. Whereas the imidazolium group of [BMIM] preferentially interacts with the pore wall in [BMIM][Cl], the cation is displaced by $FeCl_4^-$ perhaps due to the formation of a complex with the pore wall. The peak location of iron in XPS shifted downward when the IL was confined and suggests enhanced charge transfer with the pore wall. The results demonstrate the use of XPS to understand the solvent and catalytic environment of halometallate ILs of nanoconfined ILs.

537c Structure and CO_2 Capture Properties of Polyanionic Films Containing Perfluoroalkyl Sulfonylimide

Vinh Bui, Owen Lee, Leiqing Hu, Ryan Hayward, Haiqing Lin

Poly(ionic liquids) (PolyILs) with excellent mechanical stability and strong CO_2 affinity have emerged as an exciting platform for membrane carbon capture. This study aims to systematically evaluate the effect of fluorine content on structure characteristics and CO_2/gas separation properties of novel anionic PolyIL membranes. First, ionic liquid of 1-ethyl-3-methylimidazolium 3-acryloxypropylsulfonyl(-perfluoroalkyl-[sulfonimide]) with 3, 5 or 9 fluorine groups (or EF3, EF5 and EF9, respectively) were prepared via Sulfur(VI) Fluoride Exchange (SuFEx) click chemistry. These ILs can be cross-linked with 2 mol% of poly(ethylene glycol) diacrylate (PEGDA) to obtain mechanically stable PolyILs (ie. PolyEF) membrane for CO_2 separation application. Characterization of these PolyEF membranes is systematically investigated via FTIR, DSC, TGA, XRD, gas sorption and permeation apparatus. For example, increasing the flourine content between PolyEF3 and PolyEF9 increases the glass transition temperature from -39 to -12°C, and increases the counterion d-spacing from 0.43 to 0.48 nm, respectively, and enhances CO_2 permeability from 17 to 35 Barrer, and CO_2/N_2 from 10 to 17, respectively. Interestingly, free EF ionic liquid loaded PolyEFs exhibit reduction in T_g but enhancement in counterion d-spacing and simultaneous improvement in gas permeability and CO_2/gas selectivity for all samples. For instance, addition of 57wt% of EF3 in PolyEF3 decreases the T_g from -39 to -67°C, respectively, but enhances CO_2 permeability from 17 to 260 Barrer, and CO_2/N_2 from 10 to 28, respectively. Direct correlation between polymer transition temperature and gas transport properties can be satisfactorily described by the Vogel-Fulcher-Tammann (VFT) equation. This systematic study highlights the attractive CO_2/gas separation properties of polyanionic liquid membranes, providing insight into the design of polyILs membranes for practical CO_2 capture applications.

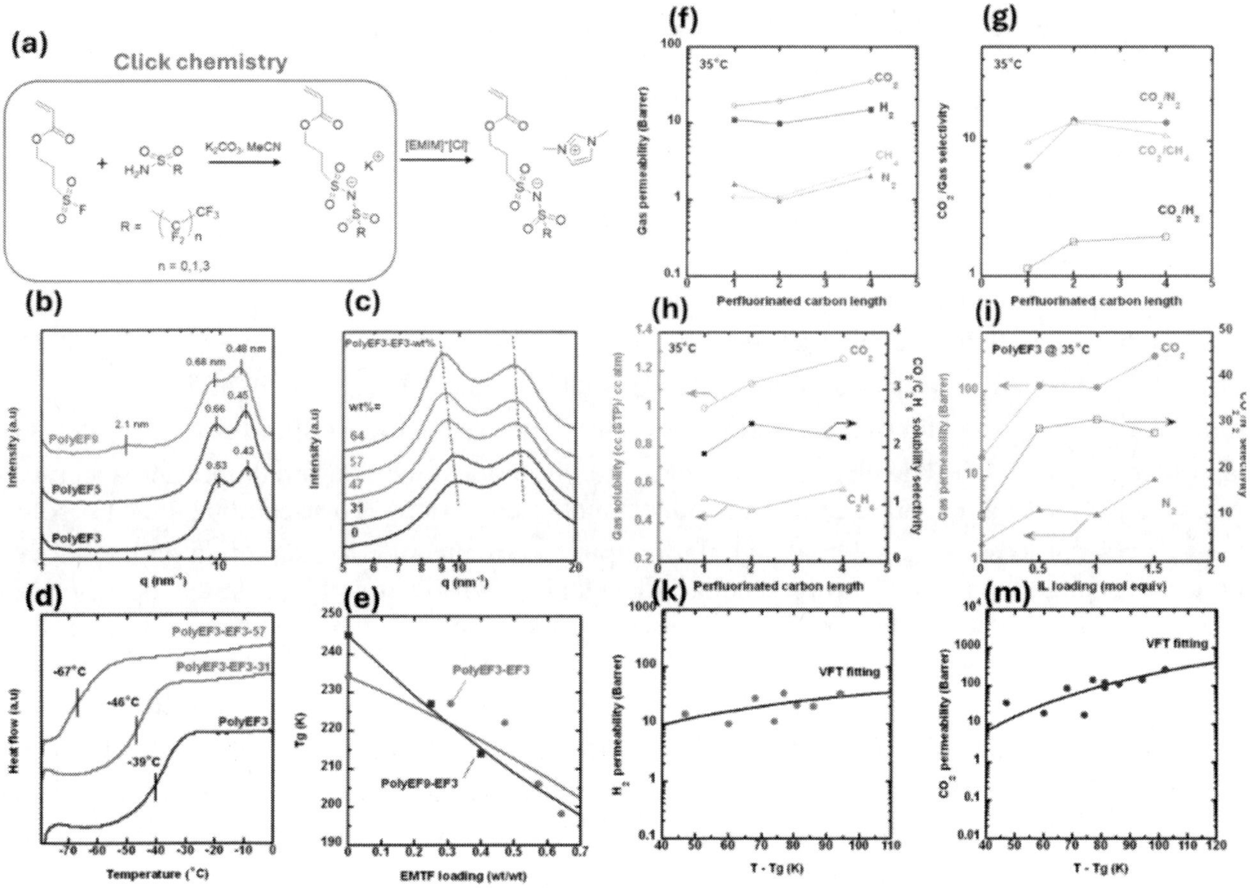
(a)
Click chemistry
K_2CO_3, MeCN
[EMIM]+[Cl]
n = 0,1,3
(b)
PolyEF9
PolyEF5
PolyEF3
q (nm⁻¹)
Intensity (a.u)
(c)
PolyEF3-EF3-wt%
wt%=
64
57
47
31
0
(d)
PolyEF3-EF3-57
PolyEF3-EF3-31
PolyEF3
-67°C
-46°C
-39°C
Heat flow (a.u)
Temperature (°C)
(e)
PolyEF3-EF3
PolyEF9-EF3
Tg (K)
EMTF loading (wt/wt)
(f)
35°C
Gas permeability (Barrer)
Perfluorinated carbon length
(g)
35°C
CO_2/Gas selectivity
Perfluorinated carbon length
(h)
35°C
Perfluorinated carbon length
(i)
PolyEF3 @ 35°C
IL loading (mol equiv)
(k)
VFT fitting
H_2 permeability (Barrer)
T - Tg (K)
(m)
VFT fitting
CO_2 permeability (Barrer)
T - Tg (K)

537d High-Throughput Electrochemical Tool to Study CO_2 Solubility in Ionic Liquids: Synergism of Void Fractions and Ionic Interactions

Rohit Chauhan, Rohan Sartape, Rashmi Mishra, Jindal Shah, Meenesh R Singh

The continuous anthropogenic emission and accumulation of CO_2 significantly contribute to climate variation associated with global warming and the greenhouse effect. Therefore, mitigating the relentless rise and accumulation of CO_2 emissions becomes a paramount global challenge. The imperative need for efficient CO_2 capture and utilization technologies to mitigate climate change has sparked interest in room-temperature ionic liquids (RTILs) as promising solvents for CO_2 capture and utilization. Most CO_2 capture processes function under pressures substantially higher than atmospheric levels, leading to increased process costs and reduced feasibility for large-scale CO_2 capture and conversion. The electrochemical method is an ideal method for simple and accurate access to CO_2 solubility values in RTILs under standard conditions. To streamline the CO_2 solubility measurement process, an automated high throughput electrochemical system utilizing the 96-wells of a microtiter plate as micro-electrolysis cells is developed, enabling quick, accurate, and sustainable assessment. This system only requires a significantly less amount (in the microliter range) of RTILs for CO_2 solubility studies. The electrocatalytic properties of electrodes and RTILs allow cyclic voltammetry (CV) to be turned into an ideal tool to determine the CO_2 solubility in RTILs with the help of a high throughput three-electrode electrochemical cell. The Cottrell analysis of the CO_2 reduction CV peak provides a direct measurement of CO_2 permeance in RTILs, which yields Henry's constant from the estimated diffusion coefficient of CO_2. Henry's constants thus obtained are in very good agreement with those reported earlier. The measured CO_2 permeance and Henry's constant of all RTILs seem to follow a first-order dependence on void fraction and a second-order dependence on electrostatic interaction between anion and cation of IL, with some synergistic dependence on the product of a void fraction and electrostatic interaction. Henry's constant decreases with increasing void fraction or decreasing electrostatic interaction. For ILs with lower void fractions, Henry's constant first increases and then decreases with increasing electrostatic interaction. Therefore, this study explores the application of CV in electrochemical measurement in a high throughput manner to determine the CO_2 solubility in RTILs. In addition, we also highlight that the influence of structural and compositional changes in RTILs on CO_2 solubility can be accurately represented by considering void fraction and electrostatic interaction and their synergistic effect, making them two significant parameters to design novel ILs.

537e Exploring Cation Effects on Transport Properties and Reaction Thermodynamics with CO_2 in Aprotic *N*-Heterocyclic Anion-Based Ionic Liquids

Pranav Thacker, Louise Canada, Joan Brennecke

In 2019 alone, 36.8 billion tonnes of carbon dioxide (CO_2) were generated, and CO_2 emissions contributed approximately 74% of total greenhouse gas emissions, resulting in the CO_2 concentration in the atmosphere reaching 421 ppm. To mitigate and control increasing CO_2 emissions, researchers have singled out carbon capture, utilization, and storage (CCSU) as a pivotal technology. This project targets post-combustion CO_2 capture from flue gas utilizing chemisorbing ionic liquids (ILs). They may be a greener alternative to aqueous amine solvents, offering negligible volatility and the ability to tune the properties of the IL for specific applications. The focus is placed on a special class of ILs with aprotic N-heterocyclic anions (AHAs) that are insensitive to undesirable viscosity increase post-CO_2 capture. By varying the anion base group, heterogeneity of substituents, and the number of ring structures, AHA ILs render enthalpic tunability and have been shown to have a prominent effect on CO_2 solubility. However, investigations of the role of the cation has been limited to phosphonium- and imidazolium-based AHA ILs. In this work, we report transport properties (i.e., density, viscosity, and ionic conductivity) and CO_2 reaction thermodynamics of triethylhexylammonium ($[N_{2226}]^+$), 1-butyl-1-methylpyrrolidinium ($[BMpyrr]^+$), and 1,1-dibutyl-2,2,3,3-tetramethylguanidin-2-ium ($[B_2TMG]^+$) cations paired with two common AHAs—2-cyanopyrrolide ($[2\text{-}CNPyr]^-$) and 3-(trifluoromethyl)pyrazolide ($[3\text{-}CF_3Pyra]^-$). Their performance will be compared to the benchmark triethyloctylphosphonium ($[P_{2228}]^+$) cation. The solubility of CO_2 will be measured at temperature of 25 °C, 40 °C, and 60 °C, with subsequent fitting of a temperature-dependent Langmuir model to determine the standard enthalpy and entropy of the reaction of the IL with CO_2 across different cation moieties. In addition, the mechanism of the reaction of these AHA ILs with CO_2 will be explored using NMR spectroscopy. A custom-build in-situ NMR (iNMR) preparation setup saturates the ILs with CO_2 inside an NMR tube, while sonication helps with the absorption kinetics. A Teflon™ valve on the NMR tube is then closed, and the tube is placed in a larger tube containing dmso-d6 (i.e., coaxial). We can then use conventional NMR instruments to determine the fraction of the total CO_2 uptake that results from reaction with the anion to form carbamate versus reaction with the cation to form carboxylates. As a result, this study widens the design space for specific applications where ILs that chemically react with CO_2 may be attractive.

537f Effect of Ionic Liquid Molar Volume on CO_2/N_2 and CO_2/CH_4 Pure and Mixed Gas Solubility Selectivity

Mariam Balogun, Matthew Davenport, Kristofer L. Gleason, Benny Freeman, Joan Brennecke

Ionic liquids (ILs) are alternative candidates for CO_2 capture due to their negligible vapor pressure, and ability to tune the cations and anions based on the application needs. This class of solvents have been evaluated extensively to capture CO_2 from post-combustion and natural gas. However, solubility data of the less soluble N_2 and CH_4 are very scarce, which makes it difficult to select the right IL that would yield the best CO_2/N_2 and CO_2/CH_4 solubility selectivity for these industrial applications. Using COSMO-RS and available literature data, we screened multiple ILs and selected eight candidates ([bmim][Tf_2N], [bmim][DCA], [bmim][TfO], [bmim][SCN], [bmim][NO_3], [bmim][TCM], [amim][DCA], [10201mim][Tf_2N]) based on estimated solubility selectivity. We measured N_2 and CH_4 solubility in these ILs, using a gravimetric apparatus up to 140 bar to reduce uncertainty, and obtain more accurate solubilities, and pure gas CO_2/N_2 and CO_2/CH_4 selectivities. ILs with small molar volumes yield higher solubility selectivity because N_2 and CH_4 solubilities are lower, presumably due to less free volume in the IL. We have also investigated mixed gas solubility selectivity for the ILs with the highest selectivities based on the pure gas measurements since real mixed gas selectivities are frequently lower than the estimates from pure gas solubilities. For this purpose, we developed a mixed gas sorption system to investigate the phase behavior of ILs with CO_2/CH_4 and CO_2/N_2 mixtures. In particular, we seek to gain a fundamental understanding on the effect of highly soluble CO_2 dissolution on the solubility of the less soluble gas (CH_4 or N_2). We have investigated CO_2/CH_4 and CO_2/N_2 mixtures at various gas compositions in ILs at partial pressures up to 30 bar at 35 °C. The mixed gas solubilities in the IL are obtained by analyzing the number of moles of desorbed gas from the IL in a 5 cm^3 sample chamber that expands into a 500 cm^3 ballast volume, with the compositions determined with gas chromatography. The real mixed gas selectivity is compared to the ideal pure gas selectivity of these gases previously measured gravimetrically and the subsequent Henry's law constants derived from these isotherms.

537g Ionic Liquid Membrane for Selective Removal of Nanoplastics from Water: Investigating the Interfacial Interactions between Polystyrene Nanoplastics and Dom

Ashish Srivastava, Milad Esfahani, Sudhir Ravula, Jason Bara

Liquid membrane stands out as an effective method for removing contaminants from water. The integration of ionic liquids (ILs), known for their significant stability, water immiscibility, and selective removal capabilities, promises to enhance the effectiveness of the separation process. Nanoplastics, recent emerging contaminants, pose challenges as current wastewater and water treatment processes lack selective and effective removal methods. The interaction between nanoplastics including polystyrene nanoplastics (PSNPs) and humic acid (HA), a complex mixture of organic compounds, presents one of the current challenges in nanoplastics removal. This study explores the use of $[C_8C_1Im]$ $[Tf_2N]$ IL as the liquid membrane for selectively removing PSNPs, focusing on unraveling the intricate interfacial interactions between PSNPs and dissolved organic matter (DOM). The removal of five different types of PSNPs with various functional groups, such as polystyrene (PS), polyvinylpyrrolidone-modified polystyrene (PS-PVP), sulfonated-polystyrene (PS-SDS), carboxylated-modified polystyrene (PS-COOH), and amine-modified polystyrene ($PS-NH_2$) were studied. The effect of HA concentration, pH, and ionic strength on the binding behavior of PS with HA was investigated. The equilibrium time obtained for the maximum removal of PSNPs with HA was 6 h. The results indicated a significant removal of 84.33% for CPS, 92.64 % for PS-SDS, and 77.67% for PS-PVP in the presence of 10 ppm HA solution. Furthermore, π-π conjugation, dipole-dipole, and ion-dipole interaction were postulated as the binding mechanisms between humic acid and PSNPs, facilitating the transportation of humic acid into the $[C_8C_1Im]$ $[Tf_2N]$.

537h Free Energy Simulations of Uranyl Extraction

Xiaoyu Wang, Michael Servis

Extractant design is fundamental to developing solvent extraction processes for actinide separations and often focuses on optimizing the direct extractant-metal interactions. However, a more detailed understanding of energetic drivers of separations beyond primary metal coordination is often lacking, including the role of solvent in the extractant phase. In this work, we use molecular dynamics simulations with umbrella sampling and find that the organic solvent can reshape the energetics of the extractant's intramolecular conformational landscape. We calculate free energy profiles of different conformations of a representative bidentate extractant, n-octyl(phenyl)-N,N-diisobutyl carbamoyl methyl phosphinoxide (CMPO), in four different solvents: dodecane, tributyl phosphate (TBP), and dry and wet ionic liquid (IL) 1-ethyl-3-methylimidazolium bis(trifluoromethylsulfonyl)imide ([EMIM][Tf_2N]). By promoting reorganization of the extractant molecule into its binding conformation, our findings suggest how particular solvents can ameliorate this energetically unfavorable step of the metal separation process. In particular, the charge alternating nanodomains formed in ILs substantially reduce the free energy penalty associated with extractant reorganization. Using alchemical free energy calculations, we find that this stabilization persists even when we explicitly include the extracted cation. We also explore the use of alchemical free energy calculations to predict the total free energy of the uranyl extraction process. By explicitly modeling the cation exchange mechanism, where the IL cation is exchanged into the aqueous phase, we can predict how choice of IL cation affects the overall free energy of extraction. These findings provide insight into the energic drivers of metal ion separations and potentially suggest a new approach to designing effective separations using a molecular-level understanding of solvent effects.

537i Molecular Structure and Solvation Regimes of Ionic Liquids in Solution and Their Effects on Measured Thermodynamic and Transport Properties

Karim Al-Barghouti, Aaron M. Scurto, Barnabas Agbodekhe, Edward Maginn

Ionic liquids have been proposed in various chemical and physical processes due to their high tunability and unique chemical behavior. Yet, knowledge of their molecular behavior in solution and its relation to observed thermodynamic and transport properties is limited. In this study, we utilize a combination of experimental methods and molecular dynamics simulations to investigate the molecular phenomena governing the non-ideal behavior observed in several measured transport properties, with an emphasis on solution thermal conductivity. We employ several spectroscopic techniques to highlight the varying structural and solvation regimes attainable in mixtures of imidazolium ILs in solution with compressed gases such as difluoromethane, pentafluoroethane, and 1,1,1,2-tetrafluoroethane as well as other molecular solutes including dichloromethane and formamide.

537j Phosphonium- and Imidazolium-Based Ionic Liquids for High-GWP Refrigerant Mixture Separation

Abdulrhman Arishi, Mark B. Shiflett

Hydrofluorocarbon (HFC) refrigerants are undergoing an 85% phase-down over the next two decades due to their high global warming potential (GWP), as mandated by the American Innovation and Manufacturing (AIM) Act and the U.S. Environmental Protection Agency (EPA). The AIM Act regulations require a transition from HFCs to the next generation of refrigerants, hydrofluoroolefins (HFOs), which have significantly lower GWP and zero ozone-depletion potential (ODP). HFOs are being commercialized as replacements for HFCs, and blends of HFO and HFC refrigerants have been identified as promising drop-in replacements for existing equipment, maintaining performance while reducing environmental impact. Several azeotropic refrigerant mixtures, such as R-450A, R-456A, R-515B, and R-516A, offer lower GWP and similar thermophysical properties to replace HFC-134a in existing systems. However, transitioning to HFO refrigerants poses challenges, particularly in recycling due to the azeotropic nature of certain HFO/HFC mixtures. Conventional distillation methods are ineffective for separating these mixtures into pure components. Extractive distillation with an ionic liquid (IL) as the entrainer has been proposed as a solution, but its efficacy depends on the selectivity and affinity of the IL towards the mixture components.

This work identified phosphonium- and imidazolium-based ILs for the separation of commercial HFO/HFC refrigerant mixtures based on ASPEN Plus simulations. In some cases, excessive operating conditions, such as 60 equilibrium stages, high operation pressure, and higher solvent to feed ratio, were required in the extractive distillation column due to the low IL selectivity. This presentation underscores the need for measuring solubility data for HFOs, the next generation of refrigerants, in ILs that can achieve high absorption capacity, selectivity, and thermal stability as this information can significantly impact the economic viability of separation processes.

689a Synthesis of Microscopic 3D Graphene for High-Performance Supercapacitors with Ultra-High Areal Capacitance

Viet Hung Pham, Congjun Wang, Yuan Gao, Jennifer Weidman, Ki-Joong Kim, Christopher Matranga

The growth of renewable energy has accelerated in recent years, in part, due to the decreasing cost of renewable energy sources and the global drive to decarbonize the energy economy. A key technology for addressing the intermittent and distributed nature of renewables is the development of efficient and reliable energy storage systems that can reduce curtailment, match electricity demand with grid capacity, and improve grid performance and reliability. Supercapacitors have attracted attention for grid-scale applications due to their high-power density, reasonable energy densities, fast charge/discharge rates, low cost, and ultra-long cycle life. Graphene is one of the most promising supercapacitor electrode materials because of its large surface area, high electrical conductivity, good chemical stability, and excellent mechanical strength. Despite graphene being considered as an ideal supercapacitor electrode material, its use in commercial devices is limited because few methods exist to produce high-quality graphene at large scale and low-cost. We report a simple method to synthesize three-dimensional (3D) graphene by graphenization of coal tar pitch with a K_2CO_3 catalyst. This produces 3D graphenes with high specific surface areas up to 2,113 m^2/g and exceptional crystallinity (Raman I_D/I_G as low as ~ 0.15). The material has an outstanding specific capacitance of 182.6 F/g at a current density of 1.0 A/g. This occurs at mass loading of 30 mg/cm^2 which is three times higher than commercial requirements, yielding an ultra-high areal capacitance of 5.48 F/cm^2. The K_2CO_3 is recycled and reused over 10 cycles with material quality and electrocapacitive performance of 3D graphene retained and verified after each cycle. The synthesis method and resulting electrocapacitive performance properties create new opportunities for using 3D graphene more broadly in practical supercapacitor devices.

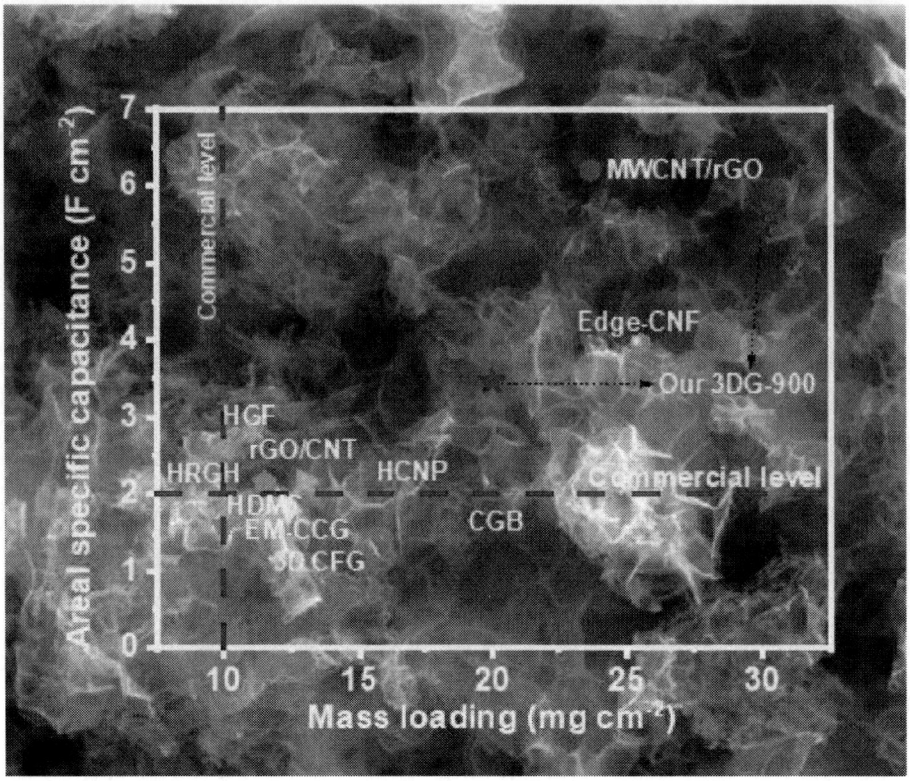
Areal specific capacitance (F cm-2)
Mass loading (mg cm-2)
Commercial level
MWCNT/rGO
Edge-CNF
Our 3DG-900
HGF
rGO/CNT
HRGH
HCNP
Commercial level
HDMC
EM-CCG
3D CFG
CGB
0
1
2
3
4
5
6
7
10
15
20
25
30

689c Diatomic Catalysts: Controlled Synthesis and Applications in Oxygen Reduction Reactions.

Helai Huang, Zhiqiang Niu

Heterogeneous catalysts containing diatomic sites are often hypothesized to have distinctive reactivity due to synergistic effects, but there are limited approaches that enable the convenient production of diatomic catalysts (DACs) with diverse metal combinations. In this report, we present a general synthetic strategy for constructing a DAC library across a wide spectrum of homonuclear (Fe_2, Co_2, Ni_2, Cu_2, Mn_2, and Pd_2) and heteronuclear (Fe–Cu, Fe–Ni, Cu–Mn, and Cu–Co) bimetal centers. This strategy is based on an encapsulation–pyrolysis approach, wherein a porous material-encapsulated macrocyclic complex mediates the structure of DACs by preserving the main body of the molecular framework during pyrolysis.

We take the oxygen reduction reaction (ORR) as an example to show that this DAC library can provide great opportunities for electrocatalyst development by unlocking unique reaction pathways. Among all investigated sites, Fe–Cu diatomic sites possess exceptional high durability for four-electron ORR because the Fe–Cu pairs can steer elementary steps in the catalytic cycle and suppress the troublesome Fenton-like reactions. We also report a diatomic cobalt catalyst for two-electron ORR to efficiently produce hydrogen peroxide at industrial-relevant current densities (200–400 mA cm^{-2}) in acid media.

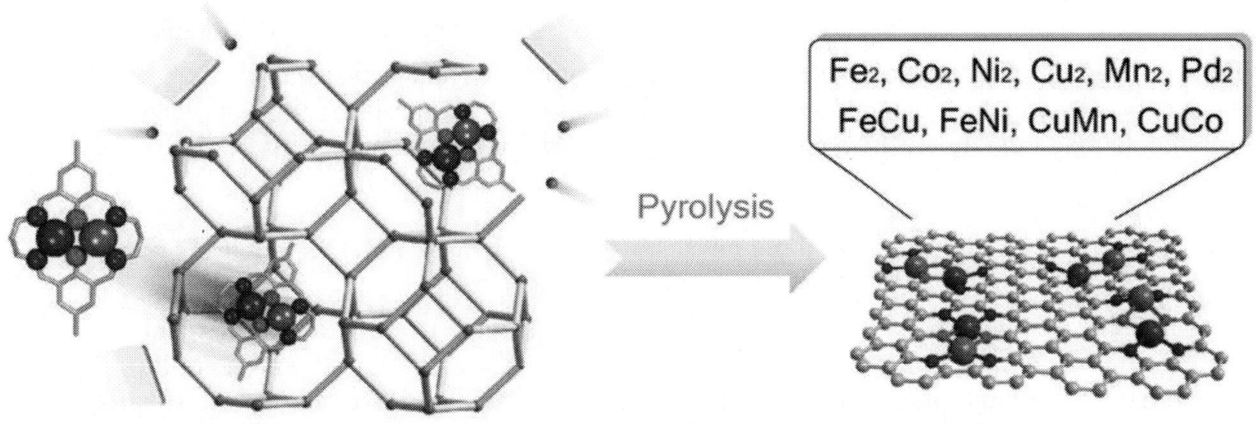

689d A Facile and Highly Efficient Method for the Construction of Mg Intercalated 1T Phase MoS_2 Materials As Anode of Li Ion Battery

Zhengrong Gu, Fuzhou Chen

Introduction:

The advancement of energy storage technologies, especially lithium-ion batteries, plays a crucial role in the shift towards renewable energy sources and the expansion of electric vehicles. Nonetheless, the search for materials that deliver higher energy densities, improved charging times, and extended durability is a constant endeavor. Transition metal dichalcogenides (TMDs), characterized by their layered structures, are significant in the domains of energy conversion and storage. Among them, molybdenum disulfide (MoS_2) stands out due to its distinct physical and chemical attributes, making it a highly promising candidate for lithium-ion battery anodes. MoS_2, through conversion reactions, possesses a theoretical capacity of 690 mAh g^{-1}, markedly surpassing the capacity of conventional graphite materials at 372 mAh g^{-1}. However, challenges such as limited conductivity and volume expansion during battery cycling impede its performance and practical application. The semiconducting nature of MoS_2 restricts charge and lithium-ion flow, leading to suboptimal utilization and slow reaction speeds. Efforts to mitigate these issues include the integration of MoS_2 with carbon or other metal oxides/sulfides to create composite materials. These heterostructures enhance local conductivity and add active sites for lithium-ion migration, although the added components may deplete electrolytes and undergo expansion, diminishing efficiency, and cycling stability.

MoS_2 exhibits polymorphism with three phases: trigonal prismatic (2H), octahedral (1T), and rhombohedral (3R), each distinguished by crystal symmetry. The 2H phase is stable but suffers from poor conductivity due to a 1.9 eV direct bandgap, whereas the 1T phase, owing to octahedral coordination of Mo and S atoms, exhibits significantly higher conductivity and metallic characteristics. Through phase engineering, the electrochemical performance of the 1T phase can be harnessed to improve MoS_2 conductivity. Nevertheless, the 1T phase's metastable nature poses challenges for its production, highlighting the importance of refining synthesis techniques. Intercalation has proven effective for creating high-purity 1T-phase MoS_2, utilizing alkaline and transition metals to stabilize the 1T phase, enhance conductivity, and expand interlayer distances. Traditional methods for synthesizing intercalated 1T-phase MoS_2, such as insertion via molten alkaline metal and exfoliation, involve complex, high-risk procedures not suited for large-scale production.

Herein, magnesium (Mg) intercalated 1T phase MoS_2 is designed and synthesized by a hydrothermal method, using regulated MBs as the precursor for the first time. The novel method can efficiently synthesize homogeneous Mg-intercalated 1 T MoS_2 with 1 T-phase purity of 90.5 % while avoiding harsh experimental conditions.

Results:

The Mg-intercalated MoS_2 demonstrated a remarkable transition to a stable 1T phase, exhibiting an enhanced surface area and providing additional active sites for lithium-ion storage. Electrochemical tests revealed a significant improvement in lithium storage capacity, achieving a higher charge storage capacity compared to conventional 2H-phase MoS_2. Moreover, the Mg-intercalated 1T-phase MoS_2 displayed superior rate capability and cycling stability, sustaining minimal capacity loss over hundreds of charge-discharge cycles.

Discussion:

The enhanced performance of Mg-intercalated 1T-phase MoS_2 is attributed to several key factors. The metallic 1T phase facilitates faster electron transfer, reducing the charge transfer resistance and enabling higher rate capabilities. Additionally, the Mg intercalation process contributes to an increased interlayer spacing, promoting faster lithium ion diffusion. This structural optimization, combined with the inherent stability of the 1T phase, underpins the outstanding electrochemical performance observed.

Conclusion:

In summary, a facile and highly efficient method for the construction of Mg intercalated high 1T phase purity MoS_2 materials is reported, using a hydrothermal method to synthesize Mg-intercalated 1T MoS_2 for the first time. Utilizing organized nanostructure MB as the precursor, the as-prepared Mg-intercalated 1T MoS_2 possesses a well-defined morphology without agglomeration and exhibits a high 1T phase purity of 90.5 %. The intercalated Mg forms an octahedral coordination with adjacent S, as shown by XPS measurement. The increased interlayer spacing created by the intercalated Mg atoms enhances the electron conductivity and electrochemical activity of the MoS_2 material. The novel Mg-intercalated 1 T MoS_2 possesses outstanding cycling and rate performance

689e N-Doped Porous Carbon Materials Produced from Covalent Organic Frameworks As Anode for Lithium-Ion Batteries

Zhengrong Gu, Jiahui Yuan

Covalent organic frameworks (COFs), a distinctive category of porous crystalline materials comprised of non-metallic elements (H, C, N, and O) interconnected by strong covalent bonds, manifesting in two-dimensional (2D) or three-dimensional (3D) structures. COFs are recognized as ideal precursors for the in-situ doping of heteroatoms into carbon matrix with uniform porosity and designable, precisely ordered porous structure. In this work, nitrogen-containing COFs were directly pyrolyzed to create nitrogen-doped porous carbon material (NPC). NPC shows high reversible charge capacities when used as the anode material in lithium-ion batteries. At current densities of 1A/g and 5A/g, NPC exhibits maximum reversible charge capacities of 436.8 mAh/g and 256.9 mAh/g, respectively. Moreover, the prepared NPC demonstrates excellent cycling stability and keeps Coulombic efficiency near 100% during the1000 cycles at 1A/g and 5000 cycles at 5A/g. These results indicated that excellent electrochemical performance of NPC is attributed to its nitrogen doping, porous structure, and high specific surface area.

689f Effects of Carbide and Oxide Filler Materials on the Ionic Conductivity of Poly(ethylene glycol) Diacrylate (PEGDA) Solid Electrolytes

Alexander Hoffmann, Wissam Fawaz, Simon Ng

Solid-state polymer electrolytes (SPEs) provide significant advantages in the production of state-of-the-art lithium metal batteries compared to conventional liquid electrolytes. SPEs are less flammable than many organic solvent based liquid electrolytes and provide resistances to many unwanted interactions in high energy battery systems. The ionic conductivities of many SPEs are substantially lower than their liquid electrolyte counterparts. The addition of filler materials is a promising method to improve the room temperature ionic conductivities of SPEs. Fillers improve the ionic conductivity of polymer electrolytes by reducing the crystallinity of the matrix or by directly participating in ionic transport. Studies on the ionic transport potential of filler materials in all-solid-state poly(ethylene glycol) diacrylate (PEGDA) electrolytes are presently limited. This study presents an in-depth comparison of ionic conductivity enhancement for various filler materials in an all-solid-state PEGDA electrolyte matrix. Through the addition of various metal oxides (Al_2O_3 and V_2O_5), metalloid oxides (SiO_2), and metal carbides (NiWC) at different concentrations, the effects of filler type and concentration on the ionic conductivity of UV-cured PEGDA electrolytes were examined with electrochemical impedance spectroscopy. Particle dispersion of the filler materials, their effects on the crystallinity and examination of the polymerization characteristics of the PEGDA were examined using SEM/EDS, XRD, and FTIR. By adding filler materials, polymer electrolyte ionic conductivities of 10^{-5} to 10^{-4} S/cm can be achieved. This study compares the benefits and drawbacks of different filler materials within a comparable polymer matrix and examines the effects that newly implemented carbide fillers have on solid-state electrolytes for the first time. Understanding the role of filler materials in the PEGDA matrix will provide insight into synthesis of high ionic conductivity solid-state electrolytes for lithium-ion and lithium-sulfur batteries as well as guide the implementation of future fillers in polymer electrolytes.

689g UV Cured Polymer Electrolyte for Solid State Lithium Sulfur Batteries

Wissam Fawaz, Alexander Hoffmann, Zhao Wang, Simon Ng

Lithium sulfur batteries (**LiS**) are heralded as the next generation of high energy battery systems due to their high theoretical specific capacity of 1675 mAh/g. Lithium sulfur batteries suffer from a loss of active sulfur which emanate from the dissolution of intermediate lithium polysulfide species during the redox reaction, resulting in diminution of cycle life. To mitigate this issue, we developed a UV cured, ionically conductive, poly (ethylene glycol) diacrylate (PEGDA) based solid polymer electrolyte. However, polymer electrolytes are plagued with a demonstratively low room temperature ionic conductivity; thus, the effects of nanoparticle fillers (silicon, silicate and silicon carbide) on the electrochemical performance were investigated. X-Ray Diffraction (XRD) was used to study the effect of the added fillers on the crystallinity of the electrolyte membrane, while Scanning Electron Microscopy (SEM) was used to understand the distribution of filler materials into the framework of the electrolyte. To gauge the electrolyte's ionic conductivity, electrochemical impedance spectroscopy (EIS) was utilized in lithium-ion blocking symmetrical cells format. Cycle life and electrochemical performance was evaluated through construction of LiS cells with a novel CoS cathode endowed with a high loading of sulfur and a lithium metal anode. Through the incorporation of lithium salt into the succinonitrile cross-linked molecules of the polymer, the electrolyte is able to deliver a room temperature, specific capacity of 296 mAh/g with improved cycle life. This study will provide a better understanding of the effects of nanoparticle fillers with PEGDA electrolytes on the energy density and cycling stability in solid state Li-S batteries.

689i Enhanced Heat Transfer for Phase Change Materials Engine in Unmanned Underwater Vehicles

Jian Liu, Wenwen Ye, Habilou Ouro-Koura, Hyunjun Jung, Adrian Garza, Aljon Salalila, Daniel Z. Deng

Phase change materials (PCM) such as paraffins have been used in the thermal engine of some unmanned underwater vehicles (UUVs) because it can provide a potential infinite energy storage solution without the limitation of battery capacities as for the current UUVs. Developing PCM and a thermal engine with enhanced heat transfer rates and increased power generation per profile for UUVs is necessary to enable more advanced sensing and ocean exploration activities. We improved the dispersion of some additives with high thermal conductivity and used them to increase the thermal conductivity of paraffin PCMs. Also, the composition of various PCM was controlled to adjust the melting temperature and heat of fusion and to achieve reduced melting time. Thermal conductivity and melting time for the PCM mixtures with additives were quantified using an off-the shelf thermal conductivity analyzer and home-made melting time test set up to correlate the relationship with the concentration of the additives, composition of PCM mixtures over the full operating temperature range. The impact of thermal conductivity and melting temperature of the PCM mixtures on the melting time were simulated using computational fluid dynamic modeling, and fin structures added to the thermal engine were proposed and evaluated to further accelerate the phase change process. Potential power generation per profile for UUVs using the modified PCM mixtures and fin structures of the PCM engine was determined based on the rate of thermal energy collection.

689j Investigation of Undoped and Strontium-Aluminum Doped Neodymium Based Perovskite with Manganese and Iron As B-Site Cation for 2-Step Thermochemical Splitting of CO2/H2O

Muhammad Usman, Khalid Al-Ali

Solar energy is considered a green alternative to mitigate the effects of fossil fuels on the environment. Among many currently accessible applications of solar energy, concentrated solar power (CSP) has advantages over solar photovoltaic (PV) for utilizing full solar spectrum & being available interminably, unlike photovoltaics which only utilize partial solar spectrum & are available only during the daytime. However, thermal energy storage systems (TES) are required in CSP's to absorb surplus solar thermal energy during the period when sunlight is available and utilize the stored energy during the night or bad weather. Various sensible, latent & thermochemical systems have been proposed for storing surplus solar thermal energy. Splitting of H_2O/CO_2 to generate the H_2/CO fuel via two step non-stoichiometric solids using cyclic thermochemical route is a promising area.

Perovskites are materials having ABO_3 type structures where A is large 12-fold coordinated alkali or alkaline earth metal cation & B is relatively small 6-fold coordinated transition metal cation forming oxygen octahedral within the perovskite crystal system. Ideal undistorted perovskite follows cubic geometry with pm3m space group. However, Orthorhombic, and rhombohedral geometries are also common in the distorted perovskite structures. Goldschmidt tolerance factor (t) relates the ionic size of the cations & anions with the possible space group of the perovskite. Doped perovskites contain structure where & are dopants. In a 2-step thermochemical cycle, the perovskites are first reduced at high temperatures using concentrated solar heat to produce non-stoichiometric oxygen deficient material (where is the oxygen non-stoichiometry. The reduced perovskites are subsequently reacted with H_2O/CO_2 to produce H_2/CO fuel simultaneously re-oxidizing the perovskite back to ABO_3 structure.

The 1st reaction (i.e. thermal reduction) is endothermic and favored at higher temperatures whereas, the 2nd reaction (i.e., re-oxidation) is exothermic and favored at low oxygen partial pressure & lower temperatures. The H_2/CO produced can be further converted to liquid hydrocarbon fuels using Fischer-Tropsch synthesis.

This research intends to report amongst the various novel perovskite compositions based on doped and undoped neodymium-manganite based perovskite the perovskite with highest yield for solar fuel (H_2/CO) production via thermochemical splitting of H_2O/CO_2. Various A site dopants are tested from Group II metals (i.e., with varying quantity except Mg where due to solubility issues. Mn and Fe were used as b-site cations doped with Aluminum where y=0, 0.2. Two-step thermochemical splitting of CO_2/H_2O is studied via redox experiments using NETZSCH thermogravimetric analyzer (TGA). Thermal reduction of perovskites was carried out at 1400 °C in 100 mL/min Argon for 45 minutes whereas re-oxidation of perovskite was carried out 800 °C in 40 %CO_2 in Argon for 60 minutes.

Perovskites are synthesized by citrate Sol-gel based modified Pechini method. The metal nitrates of the precursor cation are mixed in the desired stoichiometric molar quantities with anhydrous citric acid in 1:1.5 ratio and dissolved in 100 mL de-ionized water. Ethylene glycol is added into the mixture for polymerizing. The obtained resulting gel is dried and crushed using mortar pestle. The powder is burned at 300 °C on a hot plate and air calcined further in muffle furnace at 1400 °C for 6 hours using a ramp of 5 °C/min. Powder XRD is used to characterize the perovskite for its crystal pattern before & after Redox study using EMPYREAN PANalytical x-ray diffractometer with . ICDD database was used to compare the XRD pattrens obtained from the xrd in identification of various peak properties. Thermos scientific iCAP6000 ICP-OES is used to re-assure the metal cation ratio with the as-synthesized sample. JEOL JSM-7610F Scanning electron microscope (SEM) is used to study the pre & post redox changes in morphology of the perovskite to account for any sintering effect. The weight changes (gain or loss) of the perovskite sample are related to the oxygen non-stoichiometry of the perovskite. The oxygen non-stoichiometry data obtained through experimentation at different oxygen partial pressures is used for defect modelling of the perovskite. The research is ongoing and soon the results of the redox study will be available for publishing.

The author is interested in postdoctoral positions, any teaching or research-based opportunities and collaborations.

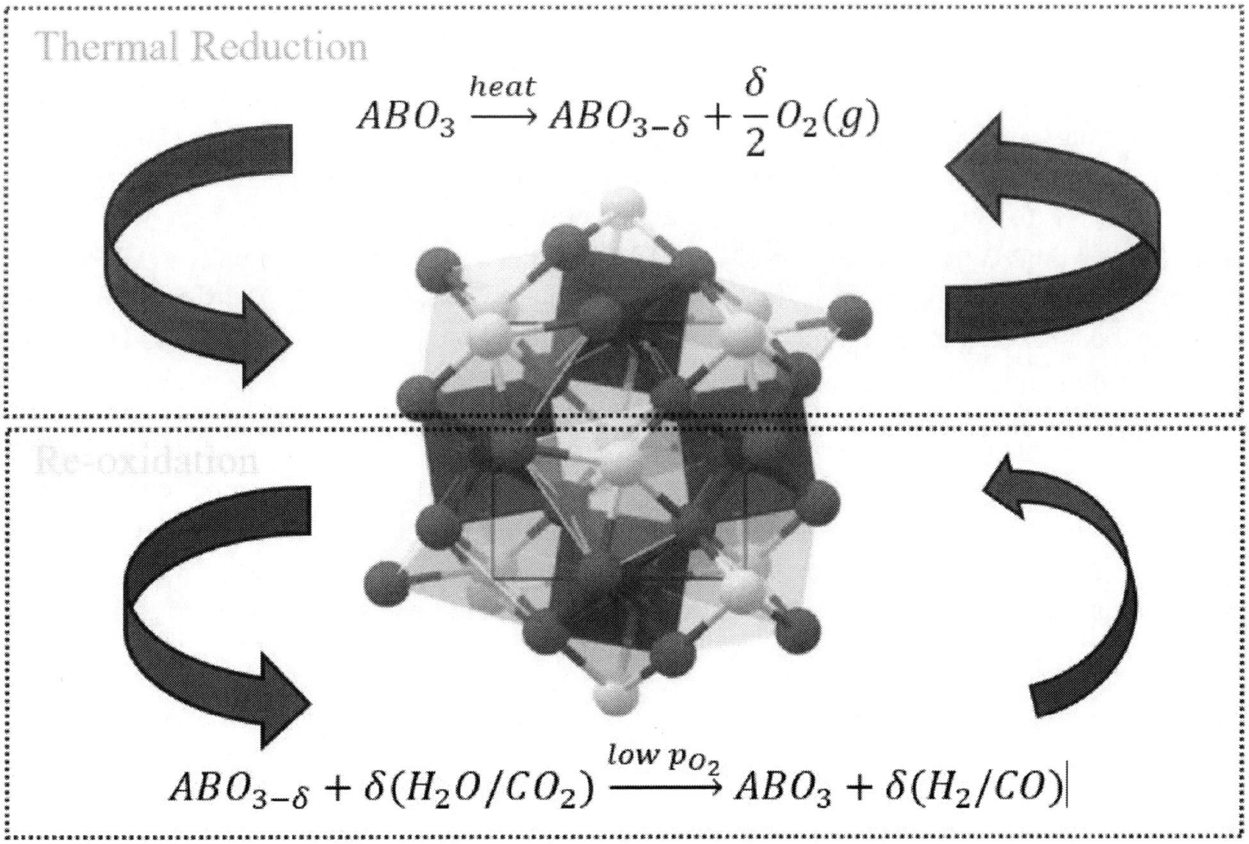

478b CO_2 Sorption Capacity and Cyclic Performance of Calcined Limestone (CaO) during Carbonation: Effects of Temperature, Steam, and Syngas (CO and H_2) Addition

Sayed Abu Sufyan, Kevin Whitty, Michael Nigra

Advanced gasification technologies like sorption-enhanced gasification featuring *in situ* carbon capture offer a promising avenue for clean H_2 production, potentially enabling the permanent sequestration of CO_2 stemming from biomass-to-H_2 using solid sorbents (e.g., lime, CaO). Despite extensive studies on CaO-based sorbents for CO_2 sorption, current studies have not clarified the effects of syngas during carbonation, and more work is required to address these effects. In this contribution, we investigate the impact of steam injection and the presence of syngas (CO and H_2) during carbonation on the sorption performance of CaO derived from commercially available limestone ($CaCO_3$). This work investigates CO_2 absorption capacity of CaO at different temperatures with or without steam injection during carbonation. Furthermore, the effects of CO and H_2 addition during carbonation are investigated in both conditions.

Our results demonstrate that steam addition positively affects the sorption activity of CaO under conditions suitable for *in situ* CO_2 capture in gasification, owing to the formation of cracks, an increase in porosity, and enhanced solid-state CO_2 diffusion through the carbonate layer. On the contrary, syngas introduction during carbonation significantly reduces the sorption performance of CaO in dry and wet conditions. It is hypothesized that competitive adsorption at CaO active sites, formation and deposition of C from the Boudouard reaction, and other competitive reactions are favorable at the same conditions as the carbonation reaction can be attributed to these inhibitory effects of syngas addition during carbonation. Furthermore, cyclic carbonation tests at these conditions are performed to understand sorbent activity and decay profiles over multiple cycles. Understanding these interactions is crucial for optimizing the performance of CaO-based sorbents for *in situ* CO_2 capture during biomass gasification.

478c Aprotic Electrochemical Systems for Continuous Carbon Capture and Utilization in Flue Gas Conditions.

Rohan Sartape, Vamsi Vikram Gande, Meenesh Singh

The exponential increase in global emission of greenhouse gases such as CO_2 is the overarching concern of the century. Considering the surge in CO_2 concentrations, there is an urgent impetus to develop novel and robust technologies that can continuously separate and capture CO_2 from flue gases. Moreover, the continuous electrochemical conversion of sequestered CO_2 to value-added products is an attractive avenue to generate revenue while mitigating emissions. Herein, we develop an aprotic system that can continuously capture and convert the CO_2 from flue gases. The developed electrochemical system can produce C_2 products at industrially realizable current densities and Faradaic Efficiencies of up to 80% under simulated flue gas conditions. To optimize the developed electrochemical system, we also explore the impact of salt concentration in electrolyte, operating voltage, and gas feed conditions. Finally, we investigate the stability of the developed electrochemical system at elevated temperatures and in the presence of impurities such as SOx to bring it one step closer to actual world application.

478d Characterization of a Pilot-Scale Supercapacitive Swing Adsorption (SSA) for Direct Air Capture of CO_2

Oluwasanmi Adeodu, Jiajie Li, Jiajie Li, Muhammad Bilal, Kai Landskron, Mayuresh Kothare

Supercapacitive swing adsorption (SSA)[1] is an electrochemical-based method of CO_2 capture and release based on charging and discharging supercapacitors. In the basic setup of SSA shown in Figure 1, a supercapacitive electrode is partially exposed to CO_2-containing gas and the other electrode is completely saturated with electrolyte. CO_2 is selectively adsorbed from the gas stream when the supercapacitor is charged, and desorption occurs upon discharge. In contrast to redox-based CO_2 sorption techniques, SSA relies on the electrostatic attraction of CO_2-derived species at the electric double layer that results from the application of a potential difference across the cell. SSA therefore has the advantage of faster charge-discharge rates, longer cycle life and higher roundtrip efficiencies over redox-based alternatives. Furthermore, the SSA process runs under mild conditions, without the additional operational costs associated with thermal and pressure swing adsorption.

Recent experimental SSA research has focused on synthesizing biomass-based supercapacitive electrodes with very high adsorption capacities[2], and characterizing the effects of charging protocols[3], electrolyte composition[4], potential oxygen interaction, on CO_2 adsorption from flue gas (15% CO_2, 85% N_2) and in direct air (400 part per million CO_2) capture (DAC) applications. Initial results suggest SSA is relatively insensitive to the choice of electrolyte, and a negative charging protocol results in a lower energy consumption than positive charging. Also, there is a short term, positive correlation between CO_2 adsorption and oxygen in the inlet gas stream. However, the concept of SSA needs to be demonstrated at pilot scale to verify its performance and for eventual scale-up for real-world commercial deployment. Furthermore, the eventual optimization of an SSA process is currently limited by the lack of clarity on the exact mechanism of CO_2 hydrolysis and potential side reactions involved in the adsorption and desorption steps.

In this work, we demonstrate an automated, scalable, four-stage semi-batch SSA process illustrated in Figure 2 for DAC. In this new architecture, the gas stream is systematically directed through valves and an extraction pump that are controlled by a Python-based logic on a central microprocessor, thus controlling the extent of CO_2 adsorption and desorption. This new experimental set-up enables the verification (or otherwise) of previous experimental findings and provides information on a previously un-examined performance metric - the purity of extracted CO_2. Experimental results and insights that are validated with this new architecture will reinforce our understanding of the SSA process, enable optimization and ultimately foster the development of commercial SSA applications for DAC.

<u>References</u>

1. Kokoszka B, Jarrah NK, Liu C, Moore DT, Landskron K. Supercapacitive swing adsorption of carbon dioxide. Angewandte Chemie International Edition. 2014;53(14):3698–3701.
2. Bilal M, Li J, Landskron K. Enhancing Supercapacitive Swing Adsorption of CO2 with Advanced Activated Carbon Electrodes. Advanced Sustainable Systems. 2023;7(11):2300250.
3. Binford TB, Mapstone G, Temprano I, Forse AC. Enhancing the capacity of supercapacitive swing adsorption CO 2 capture by tuning charging protocols. Nanoscale. 2022;14(22):7980–7984.
4. Zhu S, Li J, Toth A, Landskron K. Relationships between the Elemental Composition of Electrolytes and the Supercapacitive Swing Adsorption of CO2. ACS Applied Energy Materials. 2019;2(10):7449–7456.

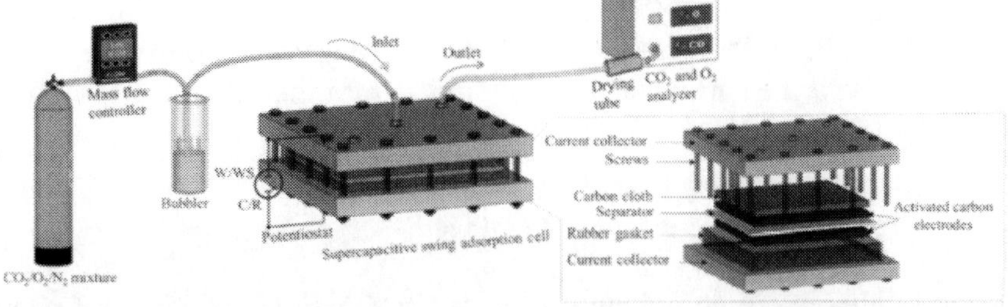

Figure 1: Experimental setup of the supercapacitive swing adsorption (SSA) process. Courtesy: Muhammad Bilal, Department of Chemistry, Lehigh University, Bethlehem, PA

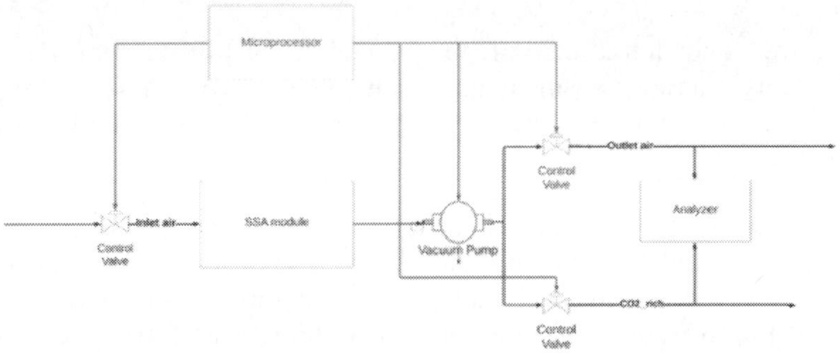

Figure 2: Automated experimental set-up. The microprocessor (Raspberry Pi) manipulates the valves and vacuum pump to regulate CO_2 adsorption and desorption. The black lines represent mass flow and the red lines represent information flow.

478e Engineering of Marine Bacteria for Carbon Sequestration By Silicate Rock Weathering

Neil Dalvie, Jan-Tobias Boehnke, Pamela Silver, Michael Springer

The Earth's oceans have capacity to permanently sequester >>1000 Gt of anthropogenic CO_2 as bicarbonate ions. The rate of marine carbon sequestration is determined by the dissolution of silicate minerals:

$$Mg_2SiO_4 + 4H_2O + 4CO_2 \text{-->} 2Mg^{2+} + H_4SiO_4 + 4HCO_3^-$$

Natural rock weathering occurs over geological time scales, however ($>10^3$ years). Enhanced weathering at large scales is precluded, therefore, by the total volume of rock that would have to be simultaneously processed to achieve short-term sequestration targets. Technologies that accelerate the rate of silicate mineral weathering above the basal rate of ~5 μm/year would enable large-scale mineral dissolution processes to capture gigatons of carbon in 10-100 years.

Recently, siderophores were shown to accelerate the dissolution of silicate minerals. These secondary metabolites are produced by microorganisms to harvest insoluble metals such as iron. Siderophores are only produced in iron-limiting conditions, however, which limits their natural abundance and subsequent catalytic effect on mineral weathering.

In this talk, we report engineering of the ubiquitous marine bacterium *Alteromonas macleodii* for accelerated weathering of the model silicate mineral olivine. We observed that petrobactin, the siderophore naturally produced by *A. macleodii*, can accelerate olivine dissolution. Then, we used simple genetic engineering tools to upregulate production of petrobactin and observed further acceleration of the mineral weathering rate (Fig 1). Finally, we characterized siderophore-mediated silicate weathering and carbonate sequestration in benchtop chemostats.

This work is the first application of synthetic biology and genome engineering for marine mineral weathering. The platform reported here will enable further optimization of strains for siderophore-mediated mineral dissolution, discovery of additional mechanisms for biological breakdown of silicates, and design of pilot scale mineral bioreactors. In the future, we envision deployment of contained, wastewater-scale, bioreactors for safe, accelerated mineral dissolution and carbon sequestration.

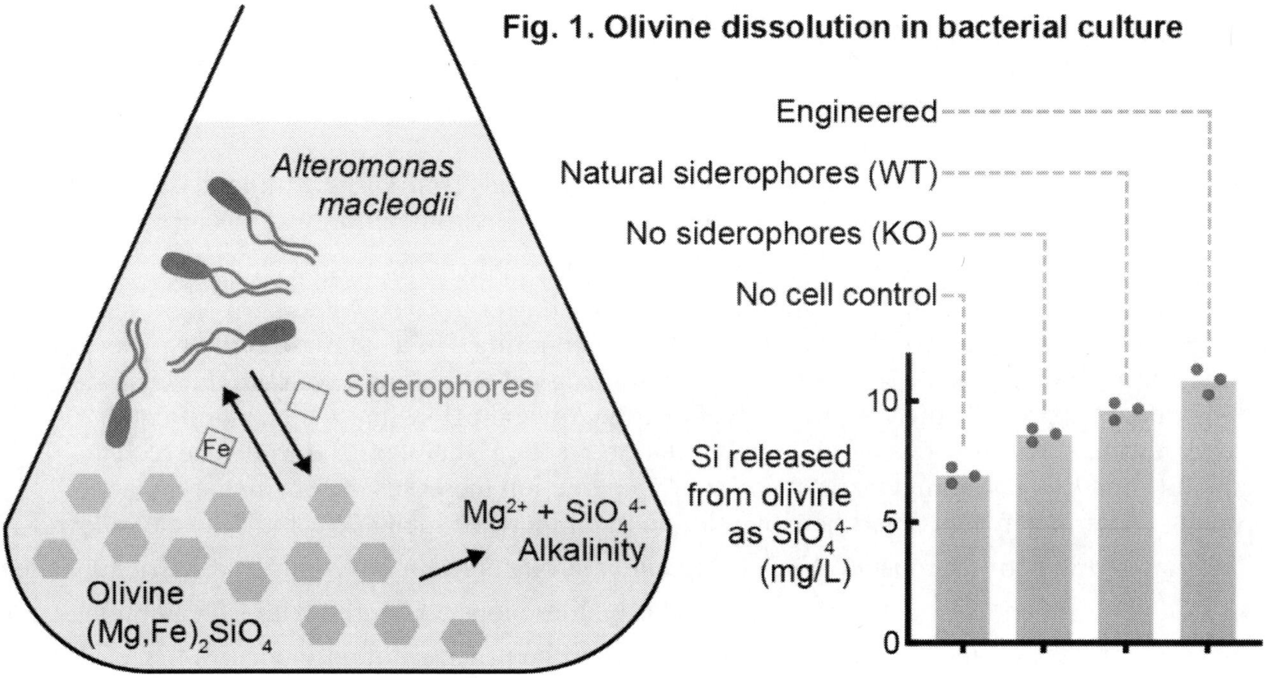

Fig. 1. Olivine dissolution in bacterial culture

478f Synergistic Metal Recovery and Carbon Mineralization from Mg-Silicate Minerals

Ning Zhang, Aaron Moment

This study presents a sustainable method for extracting metals from olivine minerals, addressing climate change and mineral scarcity. Our method employs hydrometallurgical leaching for the selective recovery of Ni^{2+} and the transformation of Mg^{2+} into carbonates, contributing to renewable energy transition and carbon sequestration. We achieved over 90% conversion of Mg^{2+} and Ni^{2+} into high purity carbonates by controlling the operational conditions of a stepwise process. We explored various inorganic/organic acids for leaching, optimizing pH levels and extracting Fe separately to obtain pure Fe salts and silica through solvent extraction and pH swing, which significantly minimizes metal loss by preventing co-precipitation reaction of silicon and iron. The techno-economic analysis confirmed that separating iron extraction increases the ecumenic impact of products. After purifying the leachate by eliminating impurities such as Fe and Si, we employed Versatic 10 to selectively separate Ni^{2+}, using either strong acids or environmentally friendly agents like CO_2 for efficient Ni^{2+} extraction. Subsequently, these ions recovered with CO_2 bubbling were converted into $NiCO_3$ through carbonation. This proposed process not only purifies the leachate, but also converts Mg into different polymorphic Mg-carbonate under ambient conditions through carbon mineralization. Our findings demonstrate effectiveness of our approach in carbon capture and utilization, and converting metals into valuable commercial products, promoting a comprehensive approach for environmental and resource management.

478g Decarbonization-Production of CO_2- Negative Building Composites

Satish Nune, David Heldebrant, Keerti Kappagantula, Yuan Jiang, Jaelynne A King, Nicholas Nelson, Jose L. Ramos, Raveen John, Reza-E-Rabby Md, Aditya Nittala, Yelin Ni, Francesca Pierobon, Ethan Nickerson, Wontae Joo

Current state of the art carbon capture and utilization studies (CCUS) approaches are centered around near-term large volume markets such as Enhanced Oil Recovery (EOR), mineralization and the production of large volumes of chemicals and fuel. These technologies have been developed to provide economic incentives to encourage the installation of CCUS technology. However, from a life cycle analysis (LCA) perspective, the products associated with these markets, such as EOR and large volume fuels, are either CO_2 neutral or, in some cases, CO_2-positive. In the intermediate term, to enable market-driven decarbonization, the US needs to develop novel materials to create new markets. This will involve the development of high value large volume products that has a long enough lifecycle time to enable the sequestration of CO_2. Despite many chemistries and markets available, CO_2-to-value-added-functional products cannot be accomplished at a scale large enough to be impactful and economically feasible. We will present a new carbon dioxide removal (CDR) approach that built upon a CO_2-derived product that is capable of sequestering hundreds of thousands of tonnes of CO_2 per year, while cost-effectively meeting commercial performance metrics against fossil-derived incumbent products. We will discuss a CO_2-negative synthetic decking material made from four large volume waste products: a recycled High-Density Polyethylene (HDPE) matrix, CO_2, and lignin or lignite fillers. We will also discuss the manufacturing process conditions with 50 -80 wt.% lignin fillers to produce composite materials. We close with a techno-economic and life cycle assessment of the envisioned capture, conversion, and composite manufacturing process.

478h Ammonia-Assisted CO2 Capture and Conversion: A Process Design and Modeling Approach for CO2 Utilization in Steelmaking

Younghun Kim, Chang Houn Rhee, Hai Woong Park, Gihun Kwon, Chi Kyu Ahn, Man Su Lee

Abstract

In response to the imperative of achieving carbon neutrality, the steelmaking industry has intensified its focus on CO2 capture and utilization technologies. This study explores the integration of ammonia-based CO2 capture and conversion within coke oven operations, leveraging attached carbon and CO2 reactions. Through comprehensive process modeling and economic evaluation, we assess the feasibility of commercial-scale implementation. Our findings indicate that this integrated approach not only enhances energy efficiency but also holds promise as a sustainable solution for CO2 utilization in the steel industry.

Introduction: The release of carbon dioxide (CO_2), a greenhouse gas in to the atmosphere is inevitable in high industrialized economy, because a major source of CO_2 comes from the combustion of fossil fuels in the steelmaking plant. To utilize CO_2 in the steel plant, CO_2 is converted into carbon monoxide (CO) by reacting attached carbon inside the coke oven. [1] (Fig 1.) The produced CO gas is transported to a power plant and recycled as fuel. A capture process is required to inject high purity CO_2 into the coke oven and convert it to CO. The POSCO Holdings completed the development of a coke oven CO_2 injection and CO conversion process with a scale of 1,875 Nm3/hr. It developed a 10-ton-per-day CO_2 capture process using ammonia water, which is inexpensive and has low renewable energy, instead of expensive amine-based absorbents. [2] (Fig 2.)

Methodology: This study employed the selective capture of CO2 from the 85,000 Nm3/hr Blast Furnace Gas (BFG) emitted from blast furnaces using ammonia water, followed by supplying the rich CO2 gas from the capture process to coke ovens (Fig. 3). Through the utilization of demo-scale operating data of the coke oven CO2 injection process and ammonia-based CO2 chemical adsorption process, a scale-up design plan and economic feasibility were evaluated using ASPEN process modeling and simulation. The reactions outlined in Table 1 were incorporated throughout the entire process, including all streams and unit operations.

Conclusion: The conversion of captured CO2 into CO not only increases the gas heating value from coke ovens but also facilitates savings in fossil fuel consumption. Simulation-based economic evaluation suggests that ammonia-based CO2 capture and CO conversion technology within coke ovens demonstrate energy efficiency and hold promise for CO2 utilization within the steelmaking industry.

Acknowledgments

This work was supported by the *The demonstration of CO_2 injection/conversion technology using Coke Ovenof the Korea Institute of Energy Technology Evaluation and Planning(KETEP) granted financial resource from the Ministry of Trade, Industry & Energy, Republic of Korea (No. 20214810100010)

References

Hai Woong Park (2022) , CO_2 utilization with coke oven, Symposium on CO_2 Capture and Utilization in Steelmaking Industry), 2022 KIChE Fall Meeting and International Symposium, October 2022. Busan BEXCO, Korea.

Man Su Lee, Kunwoo Han, Chi Kyu Ahn (2013), Progress on CO_2 Capture Pilot Plant at RIST, IEAGHG/IETS Iron & Steel Industry CCUS & Process Integration Workshop Date: 5th to 7th November 2013, Tokyo Tech Front, Tokyo Institute of Technology, Japan.

TaofangZeng, Wei BiaoFu (1996), The ratio CO/CO_2 of oxidation on a burning carbon surface. Combustion and Flame Volume 107, Issue 3, November 1996, Pages 197-210.

Pinsent, B.R.; Pearson, L.; Roughton, F.J.W. (1956); "The Kinetics of Combination of Carbon Dioxide with Hydroxide Ions", Trans. Faraday Soc. 1956, 52, 1512-1520.

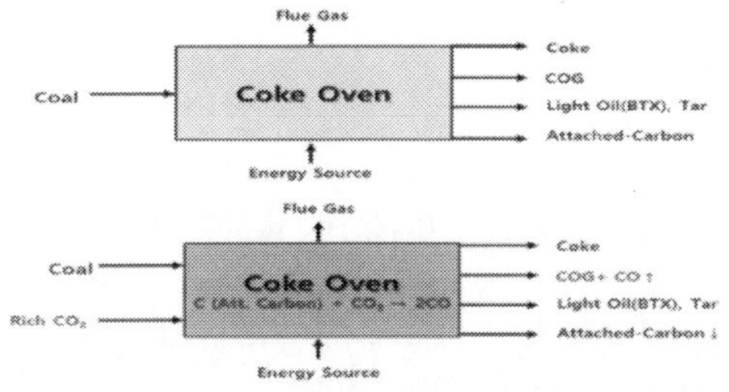

Figure 1. CO conversion process inside the coke oven (traditional/advanced)

Figure 2. 10-ton-per-day Ammonia-CO2 capture process (POSCO Holdings)

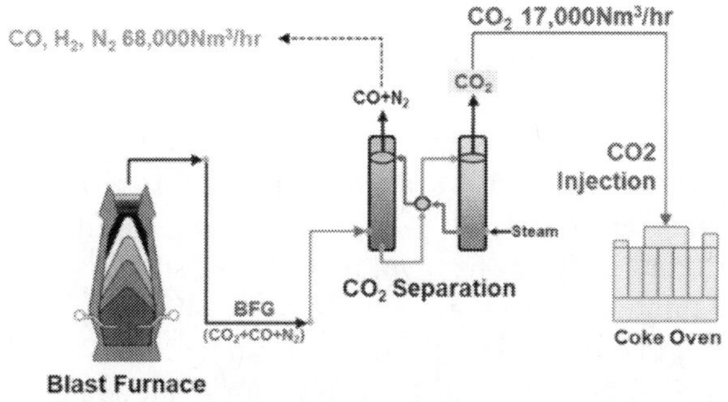

Figure 3. Ammonia-based CO_2 capture and CO conversion process

Table 1. List of reaction in the CO conversion and CO_2 capture process

Process	Reaction	Reference
CO Conversion	$C + CO_2 \rightarrow 2CO$	[3]
CO_2 Capture	$CO_2 + OH^- \rightarrow HCO_3^-$ $HCO_3^- \rightarrow CO_2 + OH^-$ $NH_3 + CO_2 + H_2O \rightarrow H_2NCOO^- + H_3O^+$ $H_2NCOO^- + H_3O^+ \rightarrow NH_3 + CO_2 + H_2O$	[4]

598a Intensification of Soybean Oil Epoxidation through Continuous Flow Millireactors

Juliana Cardenas, Alvaro Orjuela, Benjamin Katryniok, Marcia Araque Marin, Peter H. Seeberger, José Danglad-Flores

Vegetable oils are prominent feedstocks in the chemical industry, offering renewable options to fossil-based fuels, chemicals and materials. Vegetable oils can be used to produce a variety of oleochemicals including fatty acids, fatty alcohols, esters, amines, epoxides and many other widely used in many applications. The versatility of vegetable oils as feedstock relies on their composition, mainly triglycerides of saturated and unsaturated fatty acids. In particular, unsaturations in the fatty acids of triglycerides are suitable active sites for different chemical transformations including hydrogenation, halogenation, sulfonation, hydroformylation, oxidative cleave and epoxidation, etc. [1]. In particular, epoxidized vegetable oils have garnered significant attention as upgrading strategy utilized as plasticizers for polymers, imparting plasticity, fluidity, and workability to various plastic products. Given the scrutiny surrounding conventional fossil-based phthalates plasticizers due to safety concerns, the demand for biobased, biocompatible, and renewable epoxidized vegetable oils is anticipated to surge, particularly in the expanding biopolymers market. Consequently, it is imperative to explore effective pathways for their synthesis.

Traditionally, epoxidized vegetable oils have been synthesized from highly unsaturated vegetable oils, primarily soybean, linseed, and sunflower. This industrial process is mainly carried out via the Prileschajew reaction, wherein the double bonds in the vegetable oil react with hydrogen peroxide (H_2O_2), utilizing a percarboxylic acid as an intermediate and an inorganic acid as homogeneous catalyst (e.g., sulfuric acid). While this method enables high conversions, the highly exothermic nature of epoxidation and long residence times also lead to oxirane ring degradation and product polymerization, resulting in low selectivity [2], [3]. Industrially, the reaction's selectivity typically does not exceed 80%, thus limiting the maximum oxirane ring content in commercial products [4]. This highlights the need to develop intensification approaches to control reaction temperature and to increase reaction conversion, selectivity and productivity towards epoxides in a continuous process.

In this context, the use of continuous flow millireactor (Figure 1) has been evaluated in the epoxidation reaction with the aim of intensifying the process by controlling operational conditions and maximizing efficiency and selectivity. Millireactors, with their high surface-to-volume ratio, promote more efficient mass and heat transfer, resulting in shorter residence times and higher productivity compared to conventional reaction systems. In our study, various design conditions were evaluated in the millireactor, such as reactor length, stream configurations, curvature diameter of the reactor and feed conditions. Process efficiency was monitored by measuring iodine and oxirane values using hydrogen nuclear magnetic resonance spectroscopy (H-NMR). Finally, system optimization was carried out using the experimental implementation of a simplex method [5], varying operational conditions such as temperature, residence time, volumetric flow ratio of H_2O_2:oil, and volumetric flow ratio of acids (sulfuric acid and acetic acid) to oil (acids:oil). According to the obtained results, the best operating conditions were achieved at 69°C, with ratios of 0.79:1 for H_2O_2:oil and 0.29:1 for acids:oil, with a residence time of 30 minutes. These conditions resulted in a selectivity of 90% and productivity of 7.12 g/min, and an oxirane ring content of 6.4% wt. for soybean oil. These optimized operating conditions were also assessed for other vegetable oils, including canola, linseed, and sunflower with comparable yields and productivities. The obtained productivity was much larger to that observed in industrial batch processes also reducing required time for liquid-liquid separation of educts. In this regard, sets of parallel millireactors could become a suitable alternative for epoxidation of vegetable oils.

Figure 1. Scheme of the continuous slug-flow millireactor.

References

[1] G. Karmakar, P. Ghosh, K. Kohli, B. K. Sharma, and S. Z. Erhan, "Chemicals from Vegetable Oils, Fatty Derivatives, and Plant Biomass," in *ACS Symposium Series*, 2020. doi: 10.1021/bk-2020-1347.ch001.

[2] S. C. Chua, X. Xu, and Z. Guo, "Emerging sustainable technology for epoxidation directed toward plant oil-based plasticizers," *Process Biochemistry*, vol. 47, no. 10. 2012. doi: 10.1016/j.procbio.2012.05.025.

[3] P. L. Lee, W. M. Z. Wan Yunus, S. K. Yeong, D. K. Abdullah, and W. H. Lim, "Optimization of the epoxidation of methyl ester of palm fatty acid distillate," *J. Oil Palm Res.*, vol. 21, no. DECEMBER, 2009.

[4] S. Dinda, A. V. Patwardhan, V. V. Goud, and N. C. Pradhan, "Epoxidation of cottonseed oil by aqueous hydrogen peroxide catalysed by liquid inorganic acids," *Bioresour. Technol.*, vol. 99, no. 9, 2008, doi: 10.1016/j.biortech.2007.07.015.

[5] W. F. Bohórquez, A. Orjuela, P. C. Narvaez, J. G. Cadavid, J. A. García-Nunez,"Experimental optimization during epoxidation of a high-oleic palm oil using a simplex algorithm" *Ind. Crops Prod.* 187, 115321. https://doi.org/10.1016/j.indcrop.2022.115321

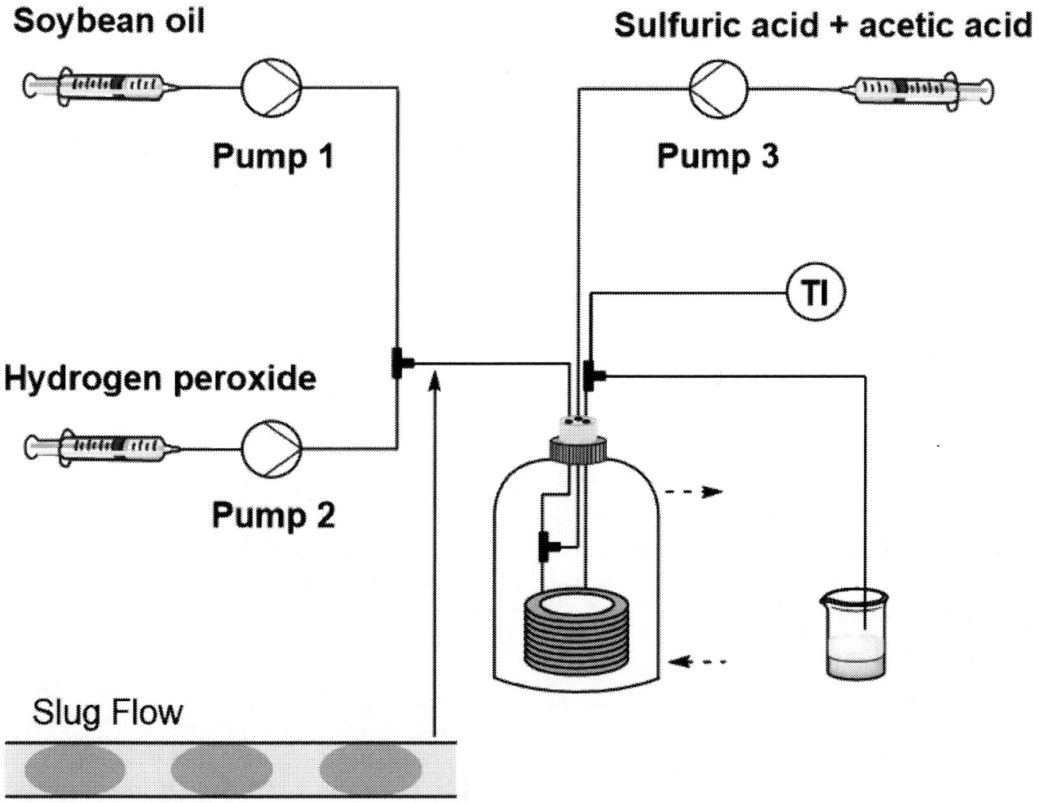

598b Process Synthesis Optimization for Microwave-Assisted Ammonia Production with Data-Driven Modeling

Md Masud, Alazar Araia, Yuxin Wang, Jianli Hu, Yuhe Tian

The pursuit of energy efficiency and environmental sustainability has stimulated the development of innovative process intensification technologies in chemical process industry. The use of microwave heating for chemical reaction synthesis presents a promising technique which can result in higher selectivity at milder reaction conditions benefited from direct and selective volumetric heating [1]. Despite the recent impetus [2-4], the modeling and simulation of microwave-assisted reaction systems remain an open challenge due to the sophisticated multi-phase (e.g., gas-solid), multi-physics (e.g., reaction, electromagnetism), and multi-scale (e.g., catalyst particle in millimeters to reactor in meters) phenomena. This also hinders to further evaluating the systems integration of these emerging technologies to retrofit existing process infrastructure [5].

To address these challenges, this work aims to develop a process synthesis optimization framework with application to microwave-assisted ammonia production. The framework features: (i) Machine learning-aided modeling of microwave reaction using experimental data, and (ii) System-level superstructure optimization of various ammonia production routes with cost and sustainability considerations. Specifically, a set of 46 data is first collected from experimental results examining the impact of reaction temperature, pressure, flowrate, and hydrogen/nitrogen ratio on ammonia concentration [6]. A neural network (NN) model is then developed using the original experimental data and the synthetic data generated via Synthetic Minority Oversampling Technique (SMOT) [7-8]. The prediction accuracy and optimization validity of the SMOTE-integrated NN model are verified against the experimental data and expert knowledge. Utilizing this data-driven model for microwave-assisted reactor, we proceed to construct a comprehensive superstructure network encompassing different commercial or emerging technology alternatives to design the overall ammonia production flowsheet. The superstructure includes options for hydrogen production through steam methane reforming and water electrolysis, as well as nitrogen extraction via air separation unit. These streams are sent to Haber-Bosch reactor and/or the microwave reactor for ammonia synthesis. The effluents from these reactors undergo separation processes, such as distillation and membrane separation, to obtain the desired ammonia product. Data-driven surrogate models are developed from first-principles simulation (e.g., Aspen Plus) using neural networks to reduce modeling complexity while ensuring sufficient physical accuracy [9]. We demonstrate the integration of these surrogate models into mixed-integer nonlinear programming for superstructure optimization, thereby systematically identifying the optimal production routes under various flowsheet design objectives (e.g., productivity, cost, sustainability).

References

[1] Hu, J., & Reddy, B. M. (Eds.). (2023). Advances in Microwave-assisted Heterogeneous Catalysis (No. 45). Royal Society of Chemistry.

[2] Goyal, H.; Chen, T. Y.; Chen, W.; Vlachos, D. G. A Review of Microwave-Assisted Process Intensified Multiphase Reactors. Chem. Eng. J. 2022, 430, 133183.

[3] Lakerveld, R.; Sturm, G.; Stankiewicz, A.; Stefanidis, G. Integrated Design of Microwave and Photocatalytic Reactors. Where Are We Now? Curr. Opin. Chem. Eng. 2014, 5, 37–41.

[4] Mevawala, C.; Bai, X.; Bhattacharyya, D.; Hu, J. Dynamic Data Reconciliation, Parameter Estimation, and Multi-Scale, Multi-Physics Modeling of the Microwave-Assisted Methane Dehydroaromatization Process. Chem. Eng. Sci. 2021, 239, 116624.

[5] Industrial Decarbonization Roadmap. Department of Energy. September 2022.

[6] Wang, Y., Khan, T. S., Wildfire, C., Shekhawat, D., & Hu, J. (2021). Microwave-enhanced catalytic ammonia synthesis under moderate pressure and temperature. Catalysis Communications, 159, 106344.

[7] Liu, X., Xu, Y., Li, J., Ong, X., Ibrahim, S. A., Buonassisi, T., & Wang, X. (2021). A robust low data solution: Dimension prediction of semiconductor nanorods. Computers & Chemical Engineering, 150, 107315.

[8] Masud, M. A. A.; Araia, A.; Wang, Y.; Hu, J.; Tian, Y., Machine Learning-Aided Process Design for Microwave-Assisted Ammonia Production. Computer-Aided Chemical Engineering. Under review

[9] Henao, C. A., & Maravelias, C. T. (2011). Surrogate-based superstructure optimization framework. *AIChE Journal*, *57*(5), 1216-1232.

598c 3D-Printed Membranes for Multiphase/Multicomponent Liquid Separation

Lei Li

A 3D-printed membrane device with multiple selectivity has been developed for multiphase/multicomponent liquid separation. Here miscible liquid mixture is separated by solid-supported ionic liquid membrane (SILM). Immiscible liquid mixture is separated by 3D-printed hydrophilic/oleophobic membrane (HOM). On demand integration of the two membranes into the device via 3D printing provides the desired multiple selectivity for a given liquid mixture. numbering scallop is enabled by mass produced interconnecting pieces similar to LEGO construction techniques. Continuous separation operation will reduce the footprint of the device. As an intensified process it will potentially replace conventional many-step operation, including distillation and extraction, and significantly lower te energy and capital cost.

598d Simulation Based Design of Simulated Moving Bed Reactor for the Production of Ethyl Chloro Acetate

Chhavi Gupta, Sanjay M. Mahajani

Simulated moving bed reactor (SMBR) is a single-unit operation apparatus with four sections, each with a set number of beds. In this process, the reaction and separation happen simultaneously with the adsorption and desorption. This makes achieving complete conversion of the limiting reactant in equilibrium-constrained reactions feasible. Its design is difficult because of the thirteen degrees of freedom, which include feed and desorbent concentrations and associated flow rates, raffinate and extract flow rates, four sections flow rates, switch time, and column dimensions. Desorption recovery from the extract and raffinate phases completes the SMBR process and primarily controls operational costs. Conversely, the capital cost is contingent upon production . Thus, the goal of this study is to use a streamlined strategy to obtain optimum output with the least amount of solvent required. Rather than combining computationally exhaustive simulation with mathematical optimization. We used a simplified approach and the reaction of interest undertaken is ethyl chloro acetate (ECA) synthesis by reacting mono-chloro acetic acid with ethanol, catalyst/adsorbent Amberlyst-15.

The solution methodology adopted is the triangle theory, a well-developed approach for simulated moving beds (SMBs) without reaction . The process employs the linear adsorption isotherm and maintains consistency on internal flow rate ratios (m1, m2, m3 and m4) to guarantee the full conversion of the limiting reaction, the purity of the extract, and the raffinate. This hypothesis is expanded to include SMB with reaction, or SMBR. Equation for the SMBR model was solved in MATLAB with appropriate boundary conditions.

The parameters m1 and m4 were chosen carefully but m2 and m3 had a range of values as per constraints. Therefore, the feasibility of m2 -m3 pairs were studied and they were plotted on triangle diagram, as shown in Figure 1. Further, the effect of various parameters was studied such as effect of residence time , flow rates of all four sections, pseudo solid velocity etc. The parameters were optimized for m2-m3 feasible pairs through different paths that gave minimum desorbent and maximum productivity.

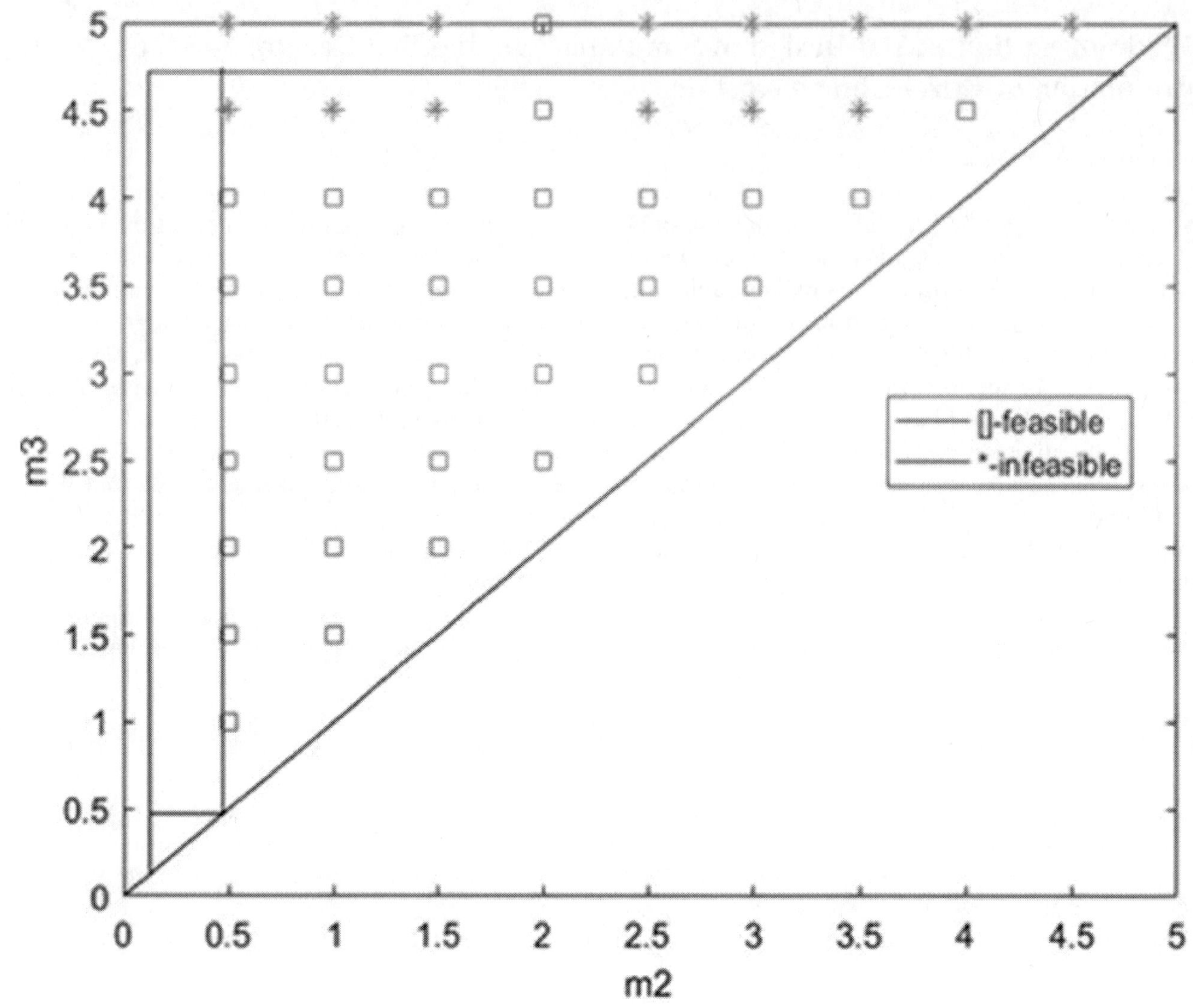

m3
m2
[]-feasible
*-infeasible

598e Hydrodynamic Cavitation-Based Leaching of Spent Cathode Material: Comparison of Orifice, Venturi and Vortex-Diode Based Cavitation Reactors

Chiara Canciani, Varaha Prasad Sarvothaman, Elia Colleoni, Paolo Guida, William Roberts

Background: The decarbonization of the transport sector and the steady increase in electrification push the demand for battery metals to unprecedented levels. The key metals for the production of lithium-ion batteries (LIB) are lithium, cobalt, nickel and manganese. The Energy Transitions Commission estimated that the potential demand of these metals is expected to spike by 2030; while the supply of lithium and nickel is forecasted to be secured by mining, the production of cobalt is expected to be sufficient to satisfy only 60% of the demand, leaving a supply gap up to 170,000 tons. Thus, the recycling of existing feedstock such as spent cathode material from end-of-life automobile batteries will be required to fulfill the deficit in demand sustainably.

Motivation: The current state of the art in recycling of spent cathode materials consists in hydrometallurgy, which uses harsh mineral acids such as H_2SO_4, and large processing times, which runs into hours for the leaching of the crushed cathode material in the acid solution. Generally, the recovery of lithium and nickel is satisfactory, while that of cobalt and manganese tends to remain low even at high acid concentrations. However, the addition of a reducing agent increases the recovery significantly, due to the change in the oxidation number of cobalt and manganese. In addition, the liquid solution is heated up to 90°C to improve the leaching rate, which posestechnical challenges in the materials selection for the equipment. Typical process time varies largely among processes, ranging from 45 minutes to 4 hours depending on the temperature, the acid concentration and the solid concentration. Therefore, leaching is a sensitive step in the process, both in terms of processing volumes, as its efficiency is inversely proportional to the solid concentration, and in terms of residence time. The leaching rate regulates the dimension of the leaching reactors, which in turn dictates the space requirements and the energy cost of the industrial facility. An intensification of the leaching rate lowers process volume and residence time, enabling the utilization of more diluted acid solutions or weaker organic acids.

Scope of this work: Cavitation - the formation, growth, and collapse of bubbles in a liquid - is a well known tool for process intensification. It generates intense shear in a confined region of the system of interest, intensifying the leaching rate. This intensification is beneficial to reduce the processing time of leaching, thus shrink the size of process equipment, and to operate at ambient conditions. To generate cavitation, the pressure of the liquid has to be lowered to a sufficient extent to form cavities. This can be achieved by applying ultrasonic waves (acoustic cavitation) or by specific fluid-flow patterns (hydrodynamic cavitation). Despite the extensive work on the applicability of acoustic cavitation, the latter is energy intensive and has limited scalability. Moreover, the lifetime of the ultrasonic probe is short due to the erosion caused by the acid and the bubble field. Being the probe a titanium alloy, its frequent substitution in a large scale scenario is resource-demanding. Compared to acoustic cavitation, hydrodynamic cavitation is less energy intensive and has good scalability potential to large process volumes. The hydrodynamic cavitation devices are less prone to erosion compared to ultrasonic probes, and are built in stainless steel. In this work, hydrodynamic cavitation was employed to increase the leaching rate, using different types of cavitation devices.

The beneficial effect of cavitation on the leaching rate is related to the enhanced species transport through the boundary layer around the cathode particles. The solid offers heterogeneous nucleation sites for the cavitation bubbles, which mainly form and collapse on the particles' surface. This improves the local transport properties, firstly favoring the penetration of the acid in the crevices on the surface of the particles, and secondly easing the transport of the leached species towards the bulk of the liquid phase. Due to the improved penetration of the acid in the particles, the average particle size is reduced by 35% compared to conventional processing techniques, for the same conversion. Therefore, the surface area exposed to the acid is higher. Being the leaching rate proportional to the exposed surface area, the latter is also higher.

Methods: To assess hydrodynamic cavitation-based leaching a cavitational rig was constructed with a centrifugal pump, pipe fittings and a pipeprimarily of 1'' size, in order to operate the different hydrodynamic cavitationdevices, namely a venturi tube, an orifice and a vortex diode. The schematic of the rig operating a vortex-diode (nominal flowrate: 20 LPM, at $\Delta P = 250$ kPag) is shown in Figure 1.

The leaching experiments were performed on NMC ($Li_{1.05}Co_{0.33}Ni_{0.33}Mn_{0.33}O_2$) cathode material, using diluted acetic acid as leaching agent and hydrogen peroxide as a reducing agent. All experiments are performed at ambient temperature. As a first step, the best cavitation device was identified, using different solid loadings. The venturi tube and the orifice are particularly sensitive to clogging when processing slurries, therefore the gradual increase in the solid loading is a key step to assess their performance. On the other hand, the vortex-diode appeared to be the best device for handling solid/liquid flows, therefore it was selected for pressure drop optimization. The pressure drop is a fundamental parameter to determine the intensity of hydrodynamic cavitation. In this work, the upstream pressure was varied from 400 to 100 kPa. Finally, the acid concentration and the hydrogen peroxide dose were optimized. In all the experiments, samples were taken at various process times to reconstruct the concentration profiles of the metals in the solution, measured by ICP-OES. The recovered solid was analyzed using SEM, to measure the particle size distribution and compare the morphological features of the particles before and after leaching.

Results: A typical leaching profile is plotted in Figure 2.The concentration of the leached metals in the acid solution reaches a plateau after 20 minutes of processing. To obtain the same conversion using the same chemicals concentration and solid loading, the observed leaching time using mechanical stirring (conventional method) is two hours.

Implications: Hydrodynamic cavitation is a technological platform relevant to the energy, water, and dairy sectors. This study expands the application horizon in the energy sector, by testing the applicability of hydrodynamic cavitation to the leaching of spent cathode material. The study reports the optimization of key reactor choices such as reactor type and operating pressure, in addition to operating parameters. The results from this study show a leaching efficiency of 75% in 15 minutes, by the use of acetic acid and hydrogen peroxide.

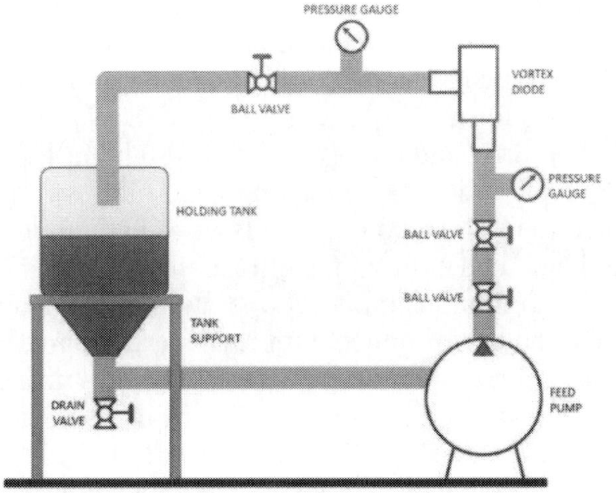

Figure 1: Schematic of the hydrodynamic cavitation rig.

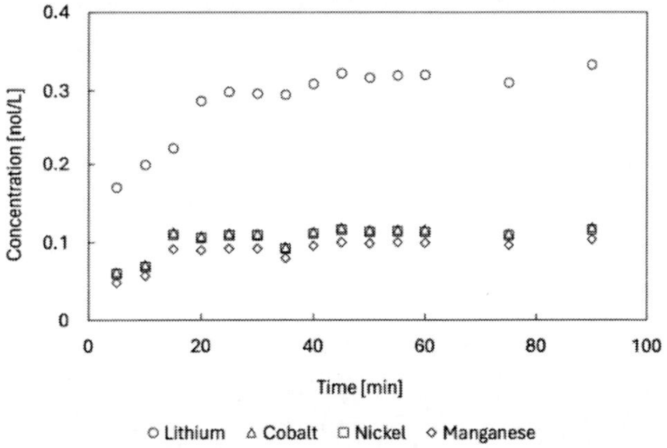

Figure 2: Concentration profile of metals leached from spent cathode (Process volume = 5-L; Solid loading = 40 g/L; H_2O/acid = 3; H_2O_2/solid loading = 0.75; HC reactor: vortex-diode; ΔP = 200 kPag).

598f Optimizing Nanoemulsion Formulation Parameters Using Continuous Ultrasonic Cavitation Technique

Diksha VATS, Vimal Kumar

Nanoemulsions (NEs) play a vital role across diverse industries, requiring efficient and scalable production methods. Ultrasonic cavitation has emerged as a promising technique for emulsion formulation due to its ability to achieve fine droplet sizes and enhance mixing efficiency. This study conducts a comprehensive comparison between batch and continuous ultrasonic cavitation methods to explain their effectiveness in NE production. Both batch and continuous modes of ultrasonic cavitation were employed in NE formulation, with a thorough evaluation of critical parameters such as droplet size distribution, stability, and energy efficiency. The influence of key process parameters including ultrasonic power, cycle ratio, residence time, and oil to surfactant ratio on emulsion characteristics was examined to understand the details of each method. Initial findings underscore the distinct advantages of both batch and continuous ultrasonic cavitation in NE formulation. In comparison with batch mode, continuous mode of operation of ultrasonic cavitation approach exhibited superior emulsification performance with less energy input, indicating its high applicability on continuous emulsification process with proper design and optimization. Optimized Ultrasonication parameters were optimized, and short- and long- term stability assessments were also conducted. At lower oil to surfactant ratio NEs was found stable after 30 days. Reduction in average droplet size (D_{avg}) was observed with increase in energy density. By understanding the details of the cavitation process and using advanced optimization strategies, this study contributes to the advancement of efficient and scalable emulsion production methods across various industrial sectors.

268a Carbon-Negative Production of Hydrogen from Biomass and Waste – Opportunities and Challenges

Kevin Whitty

Global interest and investment in hydrogen as a carbon-free energy carrier for power generation, transportation and industrial use has seen a recent upsurgence. Electrolysis based on renewable power from wind and solar is the standard approach for fossil-free production of hydrogen. However, biomass and waste can also be used to produce hydrogen with the potential to incorporate capture and sequestration of atmospheric CO_2 (via vegetation), thus achieving carbon-negative hydrogen production. Several gasification-based approaches are possible. Sorption-enhanced gasification (SEG) involves the use of two interconnected fluidized beds that circulate a mixture of bed material such as olivine and limestone/lime particles. Biomass is 80-85% converted in the steam-fluidized gasifier reactor. High-hydrogen synthesis gas is produced via a combination of steam reforming, water-gas shift, and in-situ sorption of resultant CO_2 by CaO particles. Unconverted char travels with the bed particles to the combustor reactor, where fluidizing air burns the char, thus heating up the bed particles and calcining the $CaCO_3$ to release CO_2 and regenerate CaO. Operating the combustor as an oxy-fuel reactor generates flue gas that is nearly pure CO_2. Another technology involves pressurized oxygen-blown, entrained-flow gasification of biomass or sorted MSW that has been pre-liquefied by flash pyrolysis. High carbon conversion can be achieved in a few seconds to produce a very clean syngas. Deep water-gas shift to convert the gas to H_2 and CO_2 followed by pressure swing absorption results in product streams of pure H_2 and CO_2. This presentation discusses the principles and chemistry of the two technologies, describes the designs of pilot-scale process development units at the University of Utah, and shares results of lab- and pilot-scale investigations.

268b How Molecular Modeling, AI, and a Better Assessment Process can Speed the Energy Transition

William Green

In 2015 essentially all the nations of the world decided to stabilize the climate within the next few decades. To accomplish this important societal goal, low-greenhouse alternatives will need to be developed to replace most of today's energy and fuel system, and also many chemicals/materials manufacturing processes. In some cases completely new products should be developed to replace products with large environmental impacts. Society is counting on the chemical engineering community to develop most of these new processes and products, and to provide accurate information on which proposed new processes and products merit societal investment or policy support. Because of the number and scale of the products and processes that will need to be replaced, and the relatively short timeline, this is a huge challenge.

Fortunately, computer modeling methods have been developed that can significantly speed the development of new processes and products by reducing the number of experiments required. In some cases high-throughput experiments can further reduce the time required. In this talk I will discuss some of the computer modeling methods, and highlight cases where various combinations of computational chemistry, machine learning, high-throughput experiments, and process models have drastically reduced the time required to identify a promising new product or process.

However, there are many steps between identification of a promising green product or process, and its real-world deployment at a scale which would materially affect the global climate. First, investors need to be confident the new product/process will be successful before they risk the billions of dollars needed for large scale deployment. Often there is not a compelling economic case for rapidly deploying a new product or process to replace an established technology, and sometimes existing regulations make deployment impossible. So usually some policy changes are needed to allow and support the deployment of the new technology. Policy changes can reduce the investment risk, making it much more likely that investors will provide the funds needed. Both the government and the investors look to the engineering community to provide clear assessments of the climate impact and economics of each proposed process or product, to inform their policy and investment decisions. Some ideas for how the chemical engineering community could cooperate to provide the needed reliable clear assessments will be discussed.

268c From Molecules to Modular Processes: Hydrothermal Liquefaction as a Case Study of Innovations in Renewable Fuel Production

Jillian L. Goldfarb

Without substantial government subsidies, the economic viability of large-scale thermochemical conversions of biomass to sustainable transportation fuels is tenuous at best. A profitable biorefinery will require new approaches to modular unit operations and the valorization of all byproducts. The design of such processes requires us to work against the the fundamental truths of chemical engineering: namely that systems will seek out their highest entropic state, reactions and separations are disparate unit operations, and water and oil don't mix. Together we will explore how today's advances in fundamental science will accelerate tomorrow's innovations.

For example, one of the advantages of Hydrothermal Liquefaction (HTL) over other thermochemical conversion techniques is the direct treatment of wet biomass without an energy-intensive pre-drying step. During HTL, the water's dielectric constant decreases so drastically that it acts as both a reactant and an organic solvent. However, HTL suffers from low yields of a water-insoluble biocrude phase having a high oxygen content and low heating value, as well as an aqueous phase enriched in organics. Both the biocrude and aqueous phase (which are often difficult to separate) require expensive, time consuming, material- and energy-intensive post-processing. While the literature is replete with studies on the impacts of process severity (temperature, pressure, heating rate, time), pH, (co)solvents, catalysts and feedstocks on HTL product yield and selectivity, the approach to HTL research *writ large* has been to understand the impact of process conditions on product distributions and then "back out" reaction pathways. This understanding forms the basis for catalyst selection for *in situ* or downstream upgrading of biocrude. Overcoming barriers to widespread implementation of HTL for wet waste to biofuel conversions requires a radical new approach to process design. In our laboratory, we have adopted a new view of HTL as the creation of a supersaturated solution, enabled by the "un-ordering" of water's hydrogen-bonding network. Through the design and construction of a new apparatus to measure the dielectric constant during HTL, we propose to tune the selectivity of the products and make separation and product recovery more efficient.

A complementary but often disparate literature explores management of the organic-laden process water (PW). Given the high amount of biomass-based carbon that partitions into the HTL process water (PW) we must find ways to capture and valorize the carbon to enable the commercialization of HTL as a waste management strategy. Capturing and utilizing the carbon present in the PW could represent the an impactful way to reduce the cost of biofuel production via HTL. Current approaches usually focus on anaerobic digestion; the relatively low concentration of carbon in the water requires a large reactor volume and the PW's antimicrobial activity may limit the amount of PW that can be introduced to the microbial community at any one time. The carbon could be valorized through aqueous phase processing (APP), the subject of considerable effort in the literature of late. Others suggest liquid-liquid extraction (LLE) to recover organics from the aqueous phase. However, LLE is limited because of the wide range of polarities and solubilities of the aqueous phase compounds, many of which are water soluble. Organic recoveries are low even when using copious amounts of solvent. New developments such as separation of the aqueous and hydrochar/biocrude phase by centrifugation prior to LLE or solvent extraction via supercritical CO_2 may enhance product yields.

As a society, we find ourselves at the precipice of climactic ruin. To escape this fate, we must be willing to push the boundaries of conventionally accepted process wisdom to design new approaches to renewable fuel production. In this talk, we think beyond the current literature norms of "make and characterize a fuel" to envision integrated processes for holistic carbonaceous waste valorization that teeter on the edge of thermodynamic equilibrium.

AUTHOR INDEX

AIChE
120 Wall Street, FL 23
New York, NY 10005-4020

ISBN 979-8-3313-1686-0

CURRAN ASSOCIATES INC.
proceedings.com